Steel Surfaces

ZAHNER'S ARCHITECTURAL METALS SERIES

Zahner's Architectural Metals Series offers in-depth coverage of metals used in architecture and art today. Metals in architecture are selected for their durability, strength, and resistance to weather. The metals covered in this series are used extensively in the built environments that make up our world and are also finding appeal and fascination to the artist. These heavily illustrated guides offer comprehensive coverage of how each metal is used in creating surfaces for building exteriors, interiors, and art sculpture. This series provides architects, metal fabricators and developers, design professionals, and students in architecture and design programs with a logical framework for the selection and use of metallic building materials. Forthcoming books in *Zahner's Architectural Metals Series* will include Copper, Brass, and Bronze; Steel; and Zinc surfaces.

Titles in *Zahner's Architectural Metals Series* include:

Stainless Steel Surfaces: A Guide to Alloys, Finishes, Fabrication and Maintenance in Architecture and Art

Aluminum Surfaces: A Guide to Alloys, Finishes, Fabrication and Maintenance in Architecture and Art

Copper, Brass, and Bronze Surfaces: A Guide to Alloys, Finishes, Fabrication and Maintenance in Architecture and Art

Steel Surfaces: A Guide to Alloys, Finishes, Fabrication and Maintenance in Architecture and Art

Steel Surfaces

A Guide to Alloys, Finishes, Fabrication, and Maintenance in Architecture and Art

L. William Zahner

WILEY

Published by John Wiley & Sons, Inc., Hoboken, New Jersey
Published simultaneously in Canada

Library of Congress Cataloging-in-Publication Data

Names: L. William Zahner, author.
Title: Steel surfaces : a guide to alloys, finishes, fabrication and
    maintenance in architecture and art / L. William Zahner.
Description: Hoboken, New Jersey : Wiley, 2021. | Series: Zahner's
    architectural metals series | Includes index.
Identifiers: LCCN 2020021102 (print) | LCCN 2020021103 (ebook) | ISBN
    9781119541622 (paperback ; acid-free paper) | ISBN 9781119541554 (adobe
    pdf) | ISBN 9781119541646 (epub)
Subjects: LCSH: Steel–Surfaces. | Steel–Finishing. | Architectural
    metal-work. | Art metal-work.
Classification: LCC TS320 .Z285 2021 (print) | LCC TS320 (ebook) | DDC
    672—dc23
LC record available at https://lccn.loc.gov/2020021102
LC ebook record available at https://lccn.loc.gov/2020021103

Cover Design: Wiley
Cover Image: © lior2/Getty Images

Printed in the United States of America

SKY10034982_062922

This book is in honor of David Norris. A friend and mentor.

Contents

Preface

"Know your own value."

Hank Rearden of Rearden Steel

Atlas Shrugged by Ayn Rand

Steel was one of the first metals I became acquainted with early in my career. It was different than the shiny copper the shop had stacked in a neat pile or the lightweight aluminum stored in racks up off the ground. Steel was heavy, thicker than most other sheet metals, and often possessed the intricate spangle of zinc crystals on the surface from the hot-dip galvanizing process.

Steel was treated rougher. Stored in open stacks. Often coated in a layer of slick oil. It had a smell to it—the smell of machinery, the smell of industry. Steel lacked the care and concern the other metals seemed to be regarded. It was the metal used for making patterns,[1] before the shapes would be cut in copper, aluminum, or steel's royal cousin, stainless steel. After which, they would be relegated to the scrap bin to be recycled. We used to recycle all the metals, and on occasion I would take a massive load of steel scrap to the recycling yard, only to receive less than 20 dollars. It hardly seemed worth it, but we were a company that had worked with metals through the hard times of the Depression, and you wasted nothing. Every steel packaging band was collected and recycled.

The first major projects I was involved with out of college had steel siding panels for cladding the walls of large coal-powered electrical generating plants that dotted the Missouri River. These heavy panels were roll-formed from 18-gauge cold-rolled steel sheet made by INRYCO, short for the Inland Ryerson Company, a once massively large steel producer in the United States. Inland Steel Company, founded in 1893, was one of the last integrated steel companies that turned ore into steel and into semifabricated wrought materials. Its subsidiary, Inland Building Systems, merged with Ryerson and later became INRYCO, but eventually felt the impact of foreign sources of steel and modernization and efficiencies of the mini-mills.

These steel-clad powerplants have stood for over 40 years. The paint coatings used on the steel structure and on the steel panels I worked with was of very high quality and today show little signs of deterioration—some fading with time and ultraviolet exposure, but generally the surfaces are in excellent condition.

---

[1]Up until the early 1980s, patterns were made from paper blueprints. There were no CAD–CAM files. Steel patterns would be cut to for later use to make elbows or roof jacks. The patterns were hung from hooks on the wall, and when similar items were made, the patterns would be used on the layout benches. A bit archaic in light of the technology of today.

Over time, my experience moved away from the mild carbon steels of siding and metal decking. Aluminum became the base metal for high-quality paint systems. Stainless steel, copper alloys, and later zinc were the metals for design of exposed, uncoated surfaces.

But there was this other interesting steel, a paradox of metal. It was called Corten.

Here was this strange steel that you wanted the surface to corrode. It would stain everything below it while it formed the initial rust, yet it was supposed to last as long as the stainless steels.

The company built its first major building in 1982, and the designer wanted a natural appearance. He chose Corten siding for the plant portion and alclad aluminum for the office. The siding was roll-formed by INRYCO as one of the last gasps of steel production, as INRYCO was to close its doors in 1986. The architect described his design as the new growing from the old – the old being the rusty metal surface of Corten. It is interesting that over the last 40 years, it is the Corten, the weathering steel, that has embraced the future and still appears sturdy and strong while aluminum is looking fragile and antiquated.

I have worked with weathering steel, the name used as often as the older Corten—or COR-TEN®, as the inventor and trademark holder, US Steel calls it. COR stands for corrosion resistance and TEN stands for tensile strength. Both of these characteristics and more are possessed by this amazing steel.

Nearly 25 years ago we realized the major drawback, from an aesthetic standpoint, was the time it took for the real deep color to appear. People did not want to wait and watch as their building rusted and painted the sidewalks and stonework with a red stain. The idea of preweathering this type of steel seemed to be the answer. Many an artist pushed the metal to achieve this preweathering on their sculpture using acids and wetting the surface. In addition to being hazardous to one's health and safety, this process is impractical for large projects.

After a bit of testing and trials, we came up with a process we now call *Solanum,* the Latin word for eggplant. The color of eggplant is a deep purple brown, similar to what is achieved when weathering steel reaches its point of surface equilibrium. *Solanum* sounded fitting for a metal, in tune with the great Sir Humphrey Davy, who named such metals as potassium, sodium, and, of course, aluminum.

The idea with preweathering is to control the oxide development in an environment specifically controlled for this special steel to corrode slowly and form three rich forms of oxide on the surface, similar to, but much quicker than, the color that formed after years of exposure. The staining would be contained for the most part and collected in our plant.

The weathering steels have a rich color tone that comes across as a material of the earth. Like brick, wood, or patina copper, weathering steel has a very natural, pleasing appearance once the oxidation takes root. Many of the projects shown in this book demonstrate the natural character of this amazing steel.

The steels we sometimes refer to as mild steel or carbon steel are ubiquitous in our everyday environment. Unlike the weathering steels, we do not notice them until they do begin to corrode. Otherwise, they go about their business of protecting us as we drive down the road, or hold our buildings up against the forces of gravity and wind. Once they start to corrode, they get noticed like mold on bread. The carbon steels with their beautifully rugged, dark gray-blue color require some form of protection to hold back rust from developing. It is the material of battleships and tanks, armor to be abused and to withstand abuse, but a little moisture and trouble sets in.

More and more designers are seeing the intrinsic beauty of the carbon steels. Conquering, or at least forestalling, the onset of the feeling of neglect the condition of rust can portray is paramount. Iron, the main element in steel, wants to join up with oxygen – and iron has more ways to join with oxygen than we can count on both hands. There are 16 oxide forms of iron. You want to keep the steel surface dry; water is the catalyst for this coupling with oxygen.

Conversion coatings slow down the marriage with oxygen by introducing other elements such as phosphates, sulfates, and copper selenide coatings that cling to the iron surface and form a barrier of darkened color. At the same time, these coatings can offer a unique appearance while maintaining the intrinsic beauty of steel.

Of all the metals I have written about, steel has been one of the more challenging. From an art and architecture perspective, steel has played a valuable role, but as a bit player, it is an inexpensive alternative. With new techniques of preweathering the high-strength, low-alloy (HSLA) steels, this surface is being recognized as a beautiful, natural material by designers and artists around the world. The darkening, bluing, and variegate finish one can obtain from the mild carbon steels adds an entire new array of possibilities to the design community. Easy to work with, weldable, and now with appealing surface finishing, steel is giving new value to the designer and artist.

L. William Zahner

Steel Surfaces

CHAPTER 1

Introduction

The only way to know how strong you are is to keep testing your limits.

Jor – L to Superman, the man of Steel.

IRON AND STEEL

Iron and steel have a long history with mankind.

The history of contemporary civilization is intermingled with the prowess of iron and steel. Steel is an alloy of iron with a small amount of carbon, usually less than 2%. Iron has been used throughout civilization to make useful tools and armaments. Implements made from iron were harder and could hold an edge better than any other substance known at that time. Even today, hardness and strength are compared to iron.

The atomic symbol for iron is Fe, which is shortened from the Latin word for the metal, *ferrum*, which means "firmness." Iron is element 26 on the periodic chart. See Figure 1.1.

Iron sits between manganese and cobalt and in the same line with ruthenium and osmium, two very dense elements.

Iron is the fourth most abundant element found on the Earth's outer crust behind oxygen, silicon, and aluminum, while the core of the Earth is said to be composed mostly of iron.

Iron is one of the few substances that demonstrate magnetism. *Ferromagnetism* is a term given to describe a phenomenon of a few materials to show magnetic attraction. Iron is chief among these; nickel and cobalt are two other elements that exhibit this trait. Ferromagnetism occurs in the rare earth element gadolinium and a few other compounds. Neodymium, rare earth magnets, are alloys of neodymium, iron, and boron. These exhibit a strong magnetic field.

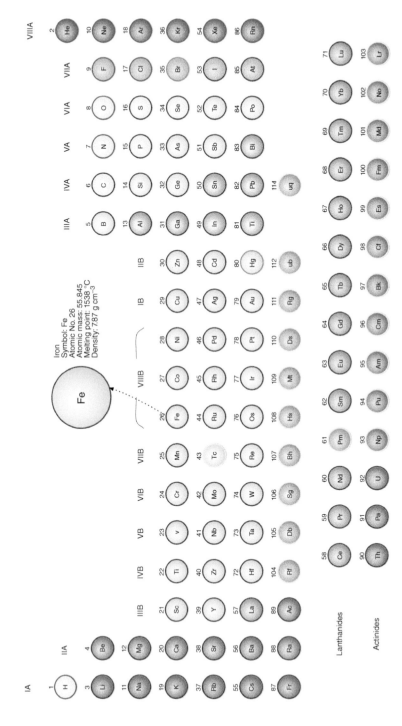

FIGURE 1.1 Periodic table of the elements.

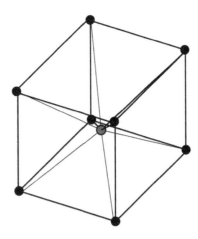

FIGURE 1.2   Body centered cubic structure of iron crystal.

One of irons chief ores, magnetite, sometimes called the *lodestone,* was known to early man as a special material that, when attached to a thread, always aligned in a given direction. This was the first compass, and the term *lodestone* means "leading stone," for it would point the way.

There is a mystery surrounding the ability of magnetite to become magnetic. Not all magnetite is magnetic, and to make it into a lodestone some strong magnetic field has to be applied. It is suggested that lightning strikes near ore deposits caused some of the magnetite to become strongly magnetized because the electrical current associated with lightning possesses a strong magnetic field and this magnetized the magnetite nearby.

Artistic adaptions to iron implements involved etching the surface of forged iron using organic acids to selectively remove areas of metal. Inlays of softer metals such as bronze, copper, and even silver could further enhance the iron surface. These softer metals could be hammered into grooves in the much harder iron, keying them into the surface to create contrasting artistic effects. Early manufacturing techniques and use of iron allowed the development of artistic surface treatments that expanded the intricate detailing already underway on the softer metals of bronze and copper.

Iron is enigmatic. With all iron's strength and hardness, it is quick to give it up. Air and moisture are all that are needed to strip this strength from iron. Those two electrons in the outer shell anxiously combine with oxygen, sulfur, or any number of other elements (Figure 1.3). Iron is never found pure in nature. Iron finds thermodynamic equilibrium when it combines with oxygen and other substances. With iron, unlike other metals such as aluminum and titanium, when it combines with oxygen and water is present, it expands as it forms oxyhydroxide. So, as the surface of iron oxidizes, it takes up more volume, creating cracks and allowing more iron under the surface to be exposed.

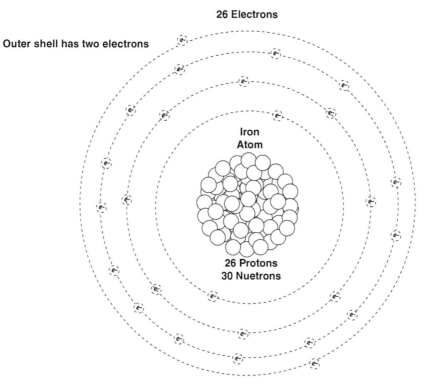

FIGURE 1.3   The iron atom.

Iron Element 26

Atomic Number 26

Crystal structure	Body-centered cube
Main mineral source	Hematite, magnetite, taconite
Color	Gray
Oxide	Black, dark red, brown, yellow
Density	7874 kg/m^3
Specific gravity	7.8
Melting point	1538 °C
Thermal conductivity	83.5 W/m °C
Coefficient of linear expansion	12×10^{-6}/°C
Electrical conductivity	17% IACS
Modulus of elasticity	200 GPa

Iron is the fourth most abundant element on Earth's surface; 6.3% of Earth's crust is composed of iron.

Steel is an alloy of iron. Carbon is the main alloying element introduced into iron to create steel. Other elements are added in small amounts to create specific properties.

Excellent ductility, deep forming ability, superior hardness, machinable. Can be both cold and hot worked.

High fracture toughness

High elasticity – resiliency under shock loading. High vibration resistance.

Hard edge. Can be sharpened and hold an edge.

Nontoxic.

Poor corrosion resistance unless alloyed with specific elements or coated with sacrificial metals.

Finishes:

Mill finish in most applications.

Rarely polished or mechanically finished.

Usually coated.

Organic coatings in the form of polymers and resins are common coatings.

Inorganic coatings in the form of glass, like porcelains, are common coatings.

Metal coatings by hot dipping or by electroplating are common coatings.

Metal oxide salts are common blackening techniques that provide both appearance and corrosion resistance.

Oxyhydroxide layers that develop on particular steel alloys called *weathering steels* are common surfaces used in art and architecture.

Artificial patina	Blacks and mottled grays. Dark greens and dark reds can be produced on steels.
Dark appearance	Iron absorbs and reflects evenly across the spectrum with slightly more on the shorter wavelengths portion of the spectrum. Alloys alter this reflection. Weathering steels have greater emission around the yellow wavelengths and the red end of the visible spectrum.

(continued)

(*continued*)

Reflectance

of Ultraviolet	Very good
of Infrared	Poor. Copper absorbs infrared wavelengths.

Relative cost	Low
Strengthening	Cold working, alloying, and tempering are methods used to adjust the strength of steels.
Recyclability	Easily recycled. Higher melting point and coatings on the steel make the scrap value very low.
Welding and joining	Can be welded, brazed, and soldered.
Casting	Steel is frequently cast using all cast methodologies.
Plating	Commonly electroplated with zinc, nickel, and chrome.
Etching and milling	Can be etched and chemically milled.

HISTORY

Iron artifacts, as old as 3500 BCE, have survived to this day. Nickel was found mixed with many of these artifacts, indicating the source of the metal was meteorites collected from the ground rather than mined. [1]

The earliest uses of iron occurred in various regions around the world. Anatonia, India, Egypt, Greece, Babylon, Japan, China, and much of northern Africa, where rich iron ore concentrations are still mined today, were some of the first regions to create iron implements for use in everyday life, warfare, and art.

Iron usage followed that of bronze, the latter being easier to refine and cast. Additionally, copper-bearing minerals were more easily identifiable due to the colorful mineral forms.

It was most likely mankind's aggressive and assertive behavior that drove the early growth and discovery of working with iron. Once mankind figured out how to cast and shape this metal, it soon supplanted copper and the copper alloy, bronze, as the material of war. The alchemist used the symbol of a diagonal arrow or the symbol for man – which is also the symbol for Mars, the god of war. See Figure 1.4.

The Hittites are said to have been one of the first civilizations to mine and work the metal by smelting ore. The Hittite civilization, also referred to as the *Kingdom of Hatti*, controlled the region around Anatolia back in 1700 BCE. They were rivals to the Egyptians. Their use of iron predates other civilizations, and because they were often at war, one can only presuppose the advantage of

---

[1] Giauque, G., 'The history of carbon steels', *The Book of Steel*, Lavoisier Publishing, 1997, p. 4.

FIGURE 1.4   Symbol of iron used by the early alchemists.

this harder and stronger material presented to the Hittites. The Chalybes were a tribe subject to the Hittites. They lived along the shores of the Black Sea. This tribe is credited with being some of the first to work with iron.

There are sites in Africa that could be even older; however, controversy surrounds their exact date. Thus, iron, entering the realm of man, is most often attributed to the central Asian region. In India, where casting and working with iron was a refined art, excavations in the Middle Ganga Valley show iron working began as early as 2800 BCE. The Mughal Empire of this era was prolific in the exploitation of iron and the early understanding of casting this metal. The metal workers of the Mughal period were experts and were some of the first to work with the lost wax technique of casting. They were known for casting near-perfect iron spheres with no seams.

It has been well established that people of this region exhibited significant prowess in the production of iron and later steel as early as 300 BCE. For the next 500 years, high-quality steel was being produced by the people of this region. They used a method referred today as the *crucible technique* to produce this high-quality steel. The steel produced was of such great quality that King Porus of India offered 15 kg of iron to Alexander the Great as a gift.

The crucible technique involved heating high-purity wrought iron mixed with charcoal and glass. The silicon in the glass would attach to impurities in the wrought iron and float to the top, while some of the carbon would be absorbed into the iron to create steel. This steel was known by the Arabs as *fülåd*, and in Europe it was called *wootz*.

The people of this region traded with the Greeks and Romans, as well as the eastern cultures. The exchange of the science of metallurgy slowly percolated out of this region to other parts of the known world.

The knowledge and ability to temper steel was well known to the metal workers of ancient India. They were known for making incredible swords and blades superior to anything at the time. Even today, some of the art of creating these special blades has yet to be uncovered.

One specialized process was the production of the *Damascus sword*, also called damascene. *Damascene* stands for the decorative process of producing wavy lines in metal by inlaying other metals or by etching the surface of metals. Figure 1.5 shows what damascene steel looked like. The Damascus sword was made from steel produced this way.

The damascene steel techniques were believed to be first developed in India. Decorative etching and metal inlay methods were perfected in India in the first millennia BCE. Romans traded with India to obtain swords and cutlery made from their specialized ironworks.

FIGURE 1.5   Damascene Steel The damascene sword techniques were believed to be first developed in India. Decorative etching and metal inlay methods were perfected in India in the first millennia BC.
Source: L. William Zahner

This early steel was known as wootz steel for the carbon content. Developed in southern India around the sixth century BCE, wootz is a steel created using this early crucible melting process. Used in India, then Damascus and the region around Toledo, Spain, this early process used sand, glass, and other substances as a flux to aid in melting iron. The resulting steel was of very high quality. Wootz is high in carbon and contained bands of pearlite, martensite, and ferrite. To make wootz, ore would be cooked inside a sealed clay crucible over a charcoal fire. Early crucible steel would use the high winds from storms to force air into the charcoal in order to achieve the necessary heat.

Little remains of early iron articles due to the propensity of iron articles to corrode when exposed to moisture and oxygen. Once corrosion would start, it was difficult to stop. There simply were no good means of inhibiting iron from wanting to combine with oxygen and form iron oxide or rust.

Most certainly, much of the early iron was collected from meteorites. This fact can be deduced from the language of the regions. The Greek word for iron is *sideros,* translated to mean, "from the stars." The Egyptian word for iron is *baaenepe,* or "gift from heaven." To melt iron, however, requires a blast furnace to achieve the temperatures needed.

Eventually, it was determined that if iron were heated for long periods of time in a crude furnace, a furnace set so that air would rush through the burning mass of charcoal and ore, a spongy lump of metal would form. This malleable iron could be hammered and flattened. Reheating would soften it or, if left for long periods of time in hot charcoal, the outer surface would harden as carbon was absorbed – processes today called *annealing* and *case hardening.*

Cast iron techniques developed out of the casting of bronze work as higher temperatures were achieved as casting processes improved. With the inclusion of small quantities of carbon, say 2–4%, the melting point of iron is reduced. Many diverse ancient cultures practiced and improved the art of casting metals. The Chinese made various small farming utensils as far back as 500 BCE, while in India exceptional casting techniques had been in development centuries earlier. Indian techniques were accepted and adopted by the Persians and the Romans as superior sources of iron products; both cast and wrought could be produced.

An example of the prowess of the Indian metallurgy is the Iron Pillar of Delhi. See Figure 1.6. This amazing structure is 8 m tall and weighs 7 tons. It was cast during the Chandragupta II reign sometime around the fourth century CE. To this day, this 1600-year-old, high-purity iron form shows little signs of corrosion.

FIGURE 1.6   Iron Pillar of Delhi, India Early Architectural Uses In 1849, the designer James Bogardus, created the first self-supporting glass and iron curtainwalls
Source: By Shutterstock

EARLY ARCHITECTURAL USES

In 1849, the designer James Bogardus created the first self-supporting glass and iron curtainwalls. These curtainwalls of iron, cast in sand molds, would stack and bolt together into multistory building fronts. The iron curtainwall proliferated in the new modern cities of eastern seaboard. In 1853, the grand Crystal Palace was erected using this new bolt-up technique of iron supporting glass walls. These intricately detailed curtainwalls of iron and glass were more economical and quicker to construct. They provided both an aesthetic appearance and a fire-resistant front to many buildings in the second half of the nineteenth century. Several major iron-casting companies sprang up across America and Europe to take advantage of the designs that could be developed using cast iron. Examples of cast iron structural columns and cast iron curtainwall are shown in Figure 1.7.

Iron is distinguished from steel in that iron is an element where steel is an alloy of iron. Iron is rarely used in its pure form. Usually, a small amount of carbon is present. Iron used in art comes in several forms, the predominate being cast and wrought. Cast is any iron that has been poured into a mold to create its form or shape, whereas wrought iron is iron that has been heated and shaped with tooling.

Figure 1.8 shows examples of early wrought iron. The upper left is a decorative sculptural form, the lower left a grille made from wrought iron, and the right image is an early clock mechanism made from wrought iron.

In the late 1700s, techniques of casting iron columns were developed and the use of iron as a structural support member proliferated. Cast iron columns provided excellent compressive strength,

FIGURE 1.7   Cast iron columns and curtainwall.
Source: L. William Zahner

FIGURE 1.8   Early wrought iron work.
Source: L. William Zahner

and designers used iron to create thin supporting storefronts, balcony supports, and even interior column supports. They were more resistant to fire than conventional wood columns and took up less space than stone supports, providing better viewing into shops with their smaller cross section. Iron columns could be cast with decorative features and could be bolted quickly into ornamental assemblies. These cast iron facades provided the needed compressive strength, stiffness, and hardness. They could not be forged or shaped because of their inherent brittleness, and catastrophe followed if they were subjected to bending and tension stresses.

Early uses of iron were of the wrought iron form. Wrought iron was in predominant use early because it did not require fully melting the ore. To fully melt the ore would require very high temperatures, so instead a hot mass of iron and other substances, usually carbon from charcoal used in the oven and slag materials from the heat, would be hammered apart to reveal small globules of iron. These globules would be collected and reheated, and then forged together by further hammering. The ability to reheat and hammer these globules of iron together to form larger elements was key to the success of iron in antiquity. Objects could be shaped this way with blows from a hammer, essentially "welding" the pieces together. The iron would join into a larger mass and would be thinned and shaped into all sorts of implements, from swords to farming implements.

Wrought iron has a long, linear grain appearance, as opposed to the regular grains of the cast irons. When broken, the grains have a fibrous or wood-like appearance. Wrought iron has lower carbon than cast iron and has inclusions from slag and oxides that are integrated by the hammering and extending of the metal grains. By integrating the slag, a more fibrous structure can be obtained, which improves flexibility.

Some telltale signs used to distinguish the difference between cast iron and wrought iron artifacts are, first, that cast iron usually is the more detailed of the two. Additionally, mold lines and detail are the domain of castings. Cast iron assemblies are bolted, whereas wrought iron assembles are riveted or heat forged together.

Wrought iron has better strength characteristics than cast iron. Wrought iron is older than cast iron because it could be created from the spongy bloom of iron in the old charcoal furnaces. These old furnaces could not reach the temperatures needed to smelt iron complete from the ore. Instead they would arrive at a carbon rich porous lump of iron mixed with various other substances. This lump would be hammered repeatedly to remove physical impurities. The larger physical impurities would spall away as the lump became thinner. Iron-rich prills could be extracted from the mass as the outer shell of glassy oxides are broken away. The prills were of rich in iron and would be hammered and folded into a larger, workable mass of wrought iron. As hammering and folding continued, the iron grains would stretch and elongate creating more strength while allowing good ductility without the brittleness of a cast iron. Wrought iron has inclusions of slag and other substances that give it a distinctive surface appearance. This grain-like structure runs through wrought iron and is created by stretching and extending these fibers through the iron as it is hammered and folded. Wrought iron has good strength, corrosion resistance and the carbon content are less. As with cast iron, today most wrought iron is actually a steel alloy, generally a low-carbon form of mild steel. Steel replaced wrought iron in the nineteenth century as new processes were introduced in the manufacture of ferrous goods.

Improvements in the manufacturing of steel started in the early 1800s. Most iron was smelted by the crucible method. This method of steel production used charcoal and later coke to produce steel from pig iron.

By the second half of the nineteenth century, the need for steel increased, and a new process, called the Bessemer process, was developed. Henry Bessemer in England and William Kelly in the United States developed a method of blasting air to change pig iron into wrought iron. This would remove carbon, but the carbon could be added back in a more controlled manner to make steel. The process is still used, and it carries the name Bessemer, who applied for a patent on the process in 1856. However, Kelly, claimed that he had also developed the process in Kentucky, in 1851. Kelly later filed for a priority claim for the process in the United States and was awarded a patent, essentially for the same process, in 1857. The Bessemer process developed because of the need of massive amounts of consistent railroad track. Prior to that, it was made from cast iron. The Bessemer process revolutionized steel making.

The commercial manufacture of wrought iron is left to small art foundries. The last major commercial foundry ceased making wrought iron in the early 1970s. Today, what we call wrought iron is actually wrought steel. Steel alloys were made more malleable and have replaced wrought iron of the past.

Forging soon developed as methods of melting iron down from other objects and roughly refined ores. Forging was an advancement over casting in that it allowed the cold working of the metal into shapes that possessed significant strength and ductility. Early iron casts were known as *iron blooms*. These were a spongy mass of iron and other impurities such as slag and charcoal that encased thin clumps of the iron. These clumps of iron would be reheated and hammered together into small shapes and utensils. These were early forms of wrought iron that contained inclusions of iron oxide and carbon from the charcoal used to heat them. These forged iron articles allowed the creation of a multitude of objects by joining clumps of metal together by hammering them while hot, essentially welding the metal together. Iron oxides, unlike copper oxides, allowed layers of iron to be joined under the heating and pressure of the early forge.

Once this occurred, iron shapes could be manufactured. Swords, helmets, and suits of armor were all hammered and shaped from forged iron. The detail and the quality of these early suits of armor are a testament to the blacksmith of old. (See Figure 1.9.)

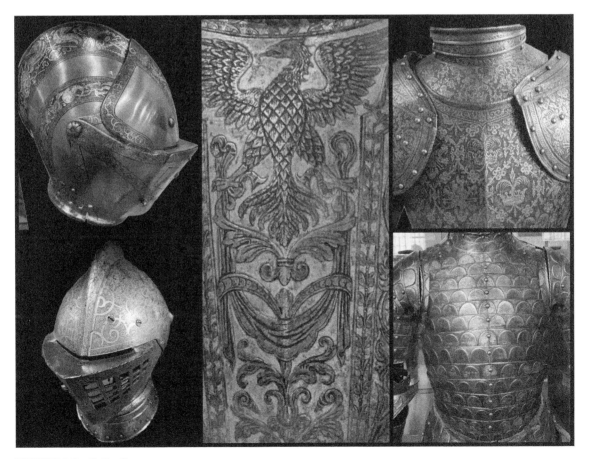

FIGURE 1.9   Suits of armor.
Source: L. William Zahner

Cast iron is a mixture of iron and more than 2% carbon with trace amounts of other elements, giving it a lower melting point than bronze. It is brittle and hard to work. Figure 1.10 show various uses of exposed cast iron in the interior of buildings. Cast iron was an excellent material choice for columns and beam supports because of the compressive strength cast iron provides.

In the nineteenth and early twentieth century, cast iron were materials considered for many common architectural embellishments. The father of American architecture, Louis Sullivan, in the early 1900s took cast iron from a utilitarian material of basic forms and shapes and created some of the most elaborate and intricate designs ever conceived (Figure 1.10). Louis Sullivan integrated cast irons of enormous complexity into his building facades and into remarkable detail elements throughout the interiors. He showed how a heavy, tough material could be made into graceful and flowing shapes that have survived for nearly 100 years (see Table 1.1).

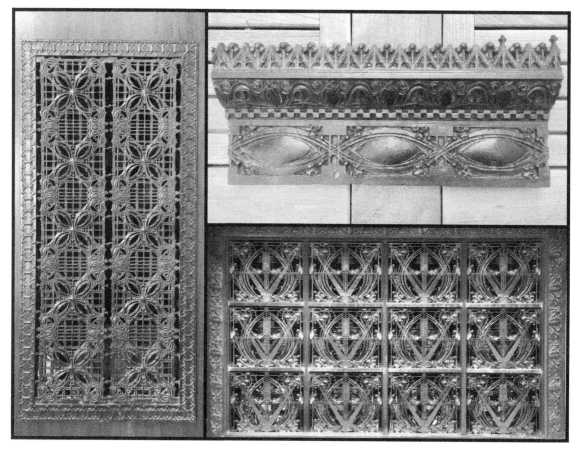

FIGURE 1.10   show various uses of exposed cast iron in the interior of buildings
Source: By Shutterstock

TABLE 1.1   Melting point of various metals.

Metal	Melting point °C	Melting point °F
Wrought iron	1482–1593	2700–2900
Steel	1371–1540	2500–2800
Cast iron	1127–1204	2060–2200
Bronze	913	1675

WHAT DEFINES STEEL

Metallurgically, steel is an alloy of iron. The term *steel* is often interchanged with iron. Not necessarily correct, but they both carry the honor of a material that is the synonym of strength and toughness. It is common for the general public to consider the alloy the same as the element.

In the wrought form, only steel is utilized. The wrought forms include sheet, plate, bar, rod, tube, pipe, wire, and hot-rolled shapes. When we think of steel, we think of the material used to create the skeletal structures that support our buildings. Figure 1.11. shows a construction site where the forms of structural steel are being assembled into a form designed to withstand environmental design loads of wind, ice, and snow.

FIGURE 1.11   Structural steel.
Source: L. William Zahner

The numerous forms and shapes of our structural steel components have specific design criteria and capacities. The steel used is special, not just any steel but a metal that has undergone testing and verification to ensure it has the appropriate mechanical properties.

Often, the most confusing aspect of steel to an artist or architect lies in the term *steel* itself. Steel is a generic term that covers a wide array of iron alloys with specific characteristics and properties. In a sense, it is like saying the term *wood* to describe the class of organic solid material made from trees.

Even the steel industry itself has a difficult time establishing a universal classification system acceptable across disciplines. Steel is so widely used in various commercial industries, across so many products and forms, it is no wonder the vocabulary used to describe steel can be so vast.

The transportation industry, in particular the automotive industry, is a major consumer of steel. However, the largest consumer of steel products is the building construction industry. Combined, nearly 70% of the steel produced goes to one of these two markets. It is no surprise the Society of Automotive Engineers (SAE) and the American Iron and Steel Institute (AISI) set out standards for the composition of various steel alloys. ASTM International, formerly the American Society for Testing and Materials, is also very instrumental in setting forth standards to assist in defining what constitutes the form and manufacture of certain steel alloys.

For the artist and architect, it is important to understand the need to work within the allowable tolerances established by many of the major users of steel. Special qualifications and requirements will be significantly more expensive if they can be achieved at all. A steel mill has processes in place that have been established for the industries it serves, and these will not be altered to meet special requirements unless it makes good business sense to them.

THE IMPORTANT STEEL ALLOYS FOR THE DESIGNER

Low-Carbon Steel

Often the term *carbon steel* is simply used for most purposes in art and architecture. Also sometimes referred to as *mild steel*, the low-carbon steels are steels that contain less than 0.30% carbon. This makes up the steel used in much of art and architecture. This steel is commonly available in wrought form and is used for structural shapes, sheets, plates, tubes, and bars and wires.

These steels are the lowest-cost steels. They are ductile, not very wear resistant in the normal supplied state; however, a process known as *carburizing* can harden the outer surface. Carburizing heats the steel in a carbon-rich atmosphere and some of the carbon is absorbed into the surface. These steels have a toughness and strength and make up the structural steels we use in building construction.

These steels will corrode in the presence of moisture or water vapor. These steels make up the base steels that are often painted or galvanized. They can be darkened. Figure 1.12 is an example of custom-darkened low-carbon steel.

FIGURE 1.12   Custom-darkened steel made from low-carbon steel (Office wall of the author.).
Source: L. William Zahner

Low-carbon steels are available in several standard alloys, both hot and cold finished. They are available in various tempers as well. These steels are easily worked, and they can be welded with all welding techniques available to the market.

Medium-Carbon Steel

These are steels that contain 0.30–0.60% carbon. These are used in the machinery industry. Hammers, rods, and axles are made of these steels. They have very high strength, as much as 620 Mpa (90 ksi). They are not commonly used in art and architecture. They do not weld as easily as the low-carbon steels and, due to their strength, are difficult to work. These steels are used for machining and forging processes within specialized industries.

High-Carbon Steel

The high-carbon steels have the highest strength of the steels, as much as 830 MPa (120 ksi). They are also known as *tool steels*. They contain 0.60–2.00% carbon. The higher carbon makes welding more difficult. These are steels used for dies, tools, and rails for the railroad. They have excellent surface hardness but lack ductility.

Tool steels are a subset of these steels. They are very strong but lack toughness and can be brittle. Saw blades and cutting tools are made from these steels as well as files and piano wire.

High-Strength, Low-Alloy Steels – The Weathering Steels

This steel is a special family of steels. They are low-carbon steels, but they have several special alloying elements added that make them atmospheric resistant once a thick, adherent oxide forms on the surface. Within this group of steels are the weathering steels. This steel type is used in art and architecture and possesses exceptional aesthetic and corrosion-resistant properties.

There are several grades of weathering steel. These steels are frequently considered for exterior art fabrications and for architectural wall surfaces. They develop a pleasing, corrosion-resistant oxide. Figure 1.13 illustrates a beautiful form of weathering steel design by Richard Serra.

Enameling Steel

These are steels with very low-carbon content. Less than 0.03%. They are used for enameling. Sometimes called *ingot iron*. The process of enameling heats the steel up to temperatures where the ceramic melts over the steel surface. If there is too much carbon the enamel will not adhere well to the surface; bubbling and pitting of the enamel will occur.

FIGURE 1.13   Sculpture by Richard Serra at The Modern, Fort Worth, Texas.
Source: Sculpture by Richard Serra

FIGURE 1.14   Porcelain on steel. Sculpture by R + K Studio.
Source: Sculpture by R + K Studio

Some enameling steel, called *decarburized steels,* has the carbon removed. Others trap the carbon by adding titanium or niobium to the alloy (Figure 1.14).

Cast Iron

Cast iron is a family of ferrous alloys that are produced by casting into forms and shapes. True cast iron is brittle. It contains large amounts of carbon, which make the surface very hard and durable. True cast iron has very good corrosion-resistance behavior. Large items, cannons, columns, and other large forms were once cast iron. Cast iron has good compressive strength but very poor tensile strength.

Cast iron alloys contain more than 2% carbon, along with other alloying elements, silicon being the main element. Usually 1–3% silicon is introduced into cast iron to benefit the casting properties.

Cast iron was once common in use for art and architecture, as shown in the work by Louis Sullivan, in Figures 1.10 and 1.15.

From an art and architectural context, it can be a "wonder through the forest" when a designer attempts to move away from some of the more common, general steel alloys.

FIGURE 1.15   Louis Sullivan cast iron designs.
Source: Verne Christensen

It is important to note that at this point, much of the steel used has a significant structural emphasis rather than an aesthetic consideration. The steel used in building construction is, for the most part, concealed from view, and therefore, the surface appearance is less important, with the exception of corrosion considerations. The importance of consistent and verified mechanical properties drives the majority of the manufacture of steel forms.

Additionally, the methods of manufacture play a significant role in choice of alloy, to the point where designations used to classify the steels include terms of manufacture such as DQ for drawing quality, SQ for structural quality, aircraft structural quality, and even saw quality for those hard steels destined for saw blade manufacture. There are many others created for the specific attributes designed for the particular industry the steel is intended. Most will rarely, if ever, fall into the realm of art or architectural surfaces.

In art and architecture, we may not be as constrained by the mechanics of one steel versus the next, but we have to appreciate these constraints and work within them.

STEEL MANUFACTURE AND PRODUCTION

The manufacturing of steel today is connected to the past. There has been and still is a tremendous evolution in the production process surrounding steel. The production of steel has evolved into many specialized industries as the efficiencies of the mini-steel mill have grown to service specific market segments. All steel production requires energy, ore, and air. This fact has never changed. Steel today is created from iron ores and scrap steel. The iron oxide minerals, hematite, and magnetite make up nearly all of the ore sources due to the large iron content these ores possess. Recycling plays a huge part in modern steel production. Steel is the most recycled of any material used by modern civilization, more than all other recycled materials combined.[2] All steel used in art and architecture today have some amount of recycled content.

Back in 1910, an International Geological Congress was held in Stockholm, Sweden, to discuss the supply of iron ore. At that meeting, it was determined the Earth's supply of steel would run out in 60 years. This even before the production of the automobile was established in force. Today, there is no worry of this occurring, as the supply of steel from recycling efforts and new ore supplies will keep steel available for a long time into the future.

Over the centuries, steel manufacturing has changed from the small cottage industry where wrought iron, the predecessor of today's steel, was hammered and quench to improve strength and hardness, crucible steel revived and modernized to create puddled steel, to massive open hearth furnaces, then to the electric arc furnace used extensively today.

The blast furnace was a technological marvel. It is a combination of improvements in steel making dating back centuries. Blast furnaces are symbols of the Industrial Age, as these gigantic structures towered over the landscape. Some of these furnaces were 40 m (130 ft) in height. They were loaded at the top and air was blasted from below. The solids would move down as the combustion particles moved skyward. The liquid metal would be drawn off and poured into what are referred to as iron pigs. They get this name because of their general shape and the way they were lined up like small pigs suckling at the sow.

Limestone was mixed with the coke to create a flux and vast amounts of carbon monoxide was generated. These giant furnaces would use more than 30 000 l (8000 gal) of water a minute. During the middle part of the twentieth century, 98% of the pig iron was produced from blast furnaces.

These pigs of refined iron still have silicon and sulfur, as well as manganese that will undergo further refinement at the next stage when they are mixed with steel scrap and enter the open hearth furnace still in use today or the more advanced electric arc furnace. This is where the steel is made as specific alloying elements are added and removed to arrive at highly controlled molten steel.

---

[2]The Steel Recycling Institute puts the recycling rate at over 85%.

The molten steel is usually further refined to remove sulfur and other impurities. The steel is degassed. Today, this molten steel is more often delivered to a continuous casting operation due to improvements in efficiencies over the ingot casting process.

Ingot casting was widely used prior to the 1950s, and terms such as *killed, semi-killed, capped,* and *rimmed steel* were used to describe the quality levels exercised in the production of ingots. Casting the ingot was the starting point for the steels. An ingot is the initial casting and was defined by a block of metal with a rectangular cross-section with rounded corners. The ingot was tapered so that one end was slightly larger in cross section than the other.

Casting involved pouring the molten metal into a vertical tapered, rectangular mold. There were always challenges to this process as the steel would solidify and shrink. A void called a *pipe* would develop, as well as pores from outgassing of carbon monoxide. Deoxidizers would be added while the steel was molten to produce *killed* steel through a process in which oxygen is removed while the metal lies quietly in the molds and little to no outgassing occurs (thus,, the term *killed*).

The mechanical properties of the killed steels are uniform and tightly defined. Their crystal structure is more uniform throughout. Structural steels and steels destined to be forged are often of this variety. Killed steel ingots often had the center void (the pipe).

The degree of removal of oxygen from the molten metal gave rise to four classes of ingot steel: killed, semi-killed, rimmed, and capped. The semi-killed kept the blowholes from the outgassing near the top of the vertical cast ingot, while the rimmed and capped ingots might have blowholes near the surfaces, leading to poor quality and spongy castings.

Today, most steel production incorporates continuous casting to eliminate the problems with ingot casting. Where ingot casting would have 20% rejection rates due to quality issues, continuous casting reduces this to 5% or less.

The continuous casting process involves the transfer of the molten metal from a reservoir such as a ladle into what is referred to as a *tundish*. The tundish operates as a control delivery system to eliminate turbulence and control feed rates. The tundish can receive several ladles in sequence, keeping the continuous casting process progressing. The tundish delivers the molten metal at a precise temperature vertically into a water-cooled mold to begin the solidification process. See Figure 1.16.

This mold is made of copper, and its shape determines the cross section of the cast. The support rolls keep the cast form moving downward as it solidifies. The speed of casting progresses at speeds of 0.5 m per minute to as much as 2.5 m per minute.

In continuous casting, the molten metal is poured continuously to make slabs for plates and sheets, blooms for structural products, and billets for round shapes such as rod, wire, and seamless pipe. Continuous casting enabled refinement and specialization in the steel industry and the efficient mini-mill operations, with less cost and improved quality, have staged a major transformation in the steel industry.

Continuous cast steel must be deoxidized or killed steel. Deoxidizers, including aluminum or silicon additives, are incorporated at the molten stage before the tundish. The need for killed steel is to eliminate porosity or blowholes at the surface.

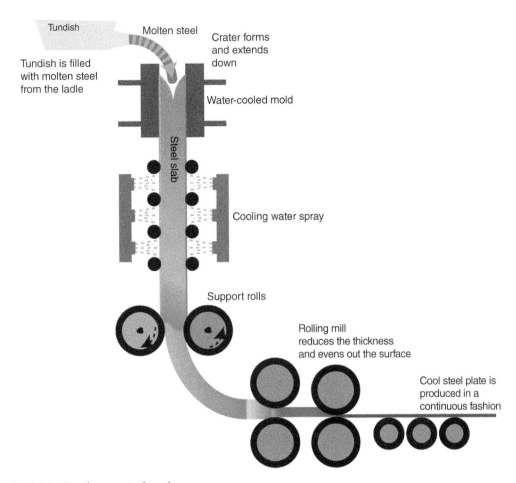

FIGURE 1.16   Continuous steel casting.

The continuous casting process of manufacture has transformed the steel industry. The large steel mill of old had to stay in continuous operation. The newer, "mini-mills" as they are sometimes called, use the electric arc furnace and scrap steel as the main source of raw materials. They can adjust easier to changes in production and demand. Figure 1.17 shows a hot, continuously cast plate.

The continuous cast method is used to produce blooms for structural shapes, billets for bars and tubes, and slabs for plates and sheets. Figure 1.18 shows the typical initial raw forms produced by continuous casting of steel.

FIGURE 1.17   Hot continuously cast steel plate.
Source: By Shutterstock

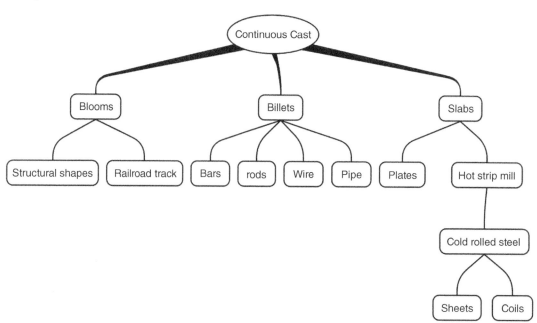

FIGURE 1.18   Product categories made from continuous casting.

SAFETY AND HYGIENE / RECYCLE

Iron is an important element for human and animal life. Iron compounds are found throughout the environment and all animal life on earth depends on some level of iron consumption. For humans and many animals, iron is a central component of hemoglobin, a protein in the blood. Humans need iron to transfer oxygen and for muscle metabolism.

Iron is a dietary supplement to ensure we get enough iron into our systems and to avoid the consequences of iron deficiencies, which can lead to fatigue, anemia, and diseases of the bone. Iron is interwoven in the health of human and animal life around the planet.

Plants need very little iron, but this small amount is necessary for plant health. Iron is used in the production of chlorophyll, which gives a plant color and is necessary for a plant to get oxygen to thrive and live.

High doses of iron, like most substances, can be toxic. Toxicity develops over time and requires the intake of large doses of iron. A human body regulates the intake of iron, but instances where too many iron supplements have been consumed and absorbed into the digestive systems have led to damage to the renal and digestive system.

Iron, like many other metals, can be recovered after the current service life has run its course. Iron can be melted and reused over and over again. This practice of recycling iron and steel dates back to the early use of the metal. Throughout antiquity, iron was gathered from the battlefield, from ship anchors and damaged cannons, to be remelted and reused.

Today the recycling of iron and steel is well established. One unique attribute of iron, its magnetism, aids in the identification and separation of iron and steel from other substances. The ease of identification has offset the low scrap value obtained in the recycling cost. Steel should never be sent to the landfill. Instead, recycling efforts gather steel from our manufacturing and building efforts and either recycle it back at the mill or repurpose it.

The repurposing of steel manufacturing scrap by Dr. Ahmed K. Ali of the Texas A&M University is a remarkable example of how to reuse steel. Figure 1.19 shows blanks from the manufacturing of cars being repurposed as a wall garden. The blanks are shaped and powder coated to create a unique, beautiful wall.

The establishment of the specialized mill with the use of the electric arc furnace has increased the demand for recycling as well as reduced the energy consumption of steel production. The energy required to recycle scrap is approximately a third of the energy required from the ore production route.[3]

The recycling of galvanized steels will recover the zinc coating and place this metal back into service.

Figure 1.20 show a galvanized steel and galvanized cast steel base that was used on a bank building in southern Kansas. Built in the 1890s, this façade "kit" was replicated over and over in small towns in the Midwest. As the useful life ends, the metal cornice will be recycled and reused on future structures.

---

[3]Beranger, G., Henry, G., Sanz, G., *The Book of Steel*, Intercept Ltd., 1996, pp. 47–49.

FIGURE 1.19   Recycled / repurposed steel sheets for a garden wall. Designed by Dr. Ahmed K. Ali, PhD, Department of Architecture, Texas A&M University.
Source: Designed by Dr Ahmed K. Ali, Ph.D. Department of Architecture, Texas A&M University

Iron production over the years has, however, taken its toll on the environment. Mining of iron ore, generally from massive strip mines, left scars on the Earth. Mining of coal to create the coke needed for steel production also has had significant negative impact on our environment, on our air and water quality. Mining has always had issues with disposal of the overburden removed to find the ore, creating dead mountains of disturbed rock and soil. Waste and impurities result from the mining operations and the dust from the operation of crushing, grinding, and separation of the ore from the waste. Vast quantities of water were used to separate the waste by floatation.

Large steel mills of old were symbols of environmental destruction with the large smokestacks spewing particulates into the atmosphere. The blast furnace towered over the horizon as these 40 m (130 ft) tall towers would combust the coke and emit large quantities of carbon monoxide combined with steam. These large steel mills would convert as much as 30,280 l (8000 gal) per minute of water into steam.

Much of the water is used in the production of coke. Coke is refined coal with a very high carbon content. The coal is heated in confined ovens away from oxygen for a period of time and just enough

FIGURE 1.20   Bank façade "kit" by the Mesker Brothers of St. Louis.
Source: L. William Zahner

to eliminate the contaminants and leave a high-purity, hard-enriched carbon behind. The refined coke is quenched in water to cool it and keep it from igniting when exposed to air.

This creates massive amounts of steam but does not release significant pollution in the form of greenhouse gases.

Today, the large steel mills around the world are being challenged and are often replaced by the smaller, more dedicated mini-mills. Since 1975, these mini-mills use 50% of the energy of the larger mills, consume less coke, and have strong connections to the recycled steel industry. These smaller mills consume scrap in the electric arc furnaces. One significant benefit of the smaller mini-mill is that they can adjust more rapidly to market fluctuations. An electric arc furnace, unlike the large blast furnaces of the integrated mills, could be shut off during times of oversupply.

Iron producers today in America, Europe, and Japan understands the need to be stewards of the environment. Reduction of energy, waste collection and recycling practices, recovery of sulfur, and scrubbing the particulates from emissions are all playing a major role. Many large operations are closed, having been replaced by the more efficient mini-mills.

Geopolitical changes have moved steel production to regions of the world where environment is often secondary. China for example, is now the largest producer of steel, nearly double the second-largest producing region. Air quality has been impacted in and around the steel-producing regions. Efforts are being made to reduce this, but steel-producing regions in China are listed as some of the most toxic air-quality regions on the planet.

COATINGS ON STEEL

In the context of art and architecture, steels are rarely used uncoated because of their tendency to absorb moisture and develop the porous and friable oxide commonly called *rust*.

Coatings can consist of a sacrificial zinc, as in the galvanizing process where the steel is electro-plated or immersed into a molten bath of zinc, as described in the bank façade shown in Figure 1.20.

Other metal coatings in common use are the aluminum–zinc coatings and Galfan™, a combination of zinc and rare earth metals. These are all considered sacrificial coatings of other metals that involve the protection of the base steel by offering up their electrons as well as acting as barrier coatings.

There are also treatments that induce a blue or black protective coating on the steel. These involve the integration of metal salts onto the surface of the steel. There is some diffusion of the iron into these coatings to create the dark, sulfide, phosphate or selenide salt. Not as protective as the sacrificial coating of zinc in the galvanized process, these coatings are often applied to small items and steel tools.

Other common coatings are the organic coatings, various resins of organic compounds. This occurs as well over the zinc coatings and phosphate conversion coatings. Figure 1.21 shows a stair-case for the design firm Kieran Timberlake, made from low-carbon steel welded and formed, and then followed by coating with an epoxy primer and finish coat of acrylic.

Additionally, inorganic coatings, created by fusing glass over the surface, known as porcelain, are excellent barrier coatings. These lack any sacrificial protection and can corrode on the edges.

One exception to this requirement for a coating is weathering steel, often referred to by the trade name COR-TEN®, or more generally as Corten,[4] is a copper-bearing steel that forms a distinctive oxide barrier over the surface. This barrier grows outward and inward from the surface. Unlike common rust, the oxide formed on weathering steel is a thick, impervious ferro-oxyhydroxide that will afford the base metal extensive resistance to corrosion. This incredible surface will be discussed extensively both from a corrosion resistant standpoint and from an aesthetic context.

---

[4]Corten or CorTen was developed by the US Steel Corporation. The name COR-TEN® is a registered trademark for the particular type of steel. US Steel developed the steel back in the 1930s. Several alloys of various compositions were developed, more for toughness and durability than for the corrosion-resistance characteristic. US Steel needed a strong steel for the hopper cars that would transfer ore and coal for steel production. However, the company discovered that this very strong and hard metal also showed good corrosion resistance, thus the name *Cor* for corrosion resistance and *Ten* for tensile strength.

FIGURE 1.21   Kieran Timberlake designed steel staircase.

SURFACE COATINGS OF STEEL USED IN ART AND ARCHITECTURE

- Coated with other metals – zinc, zinc–aluminum, chrome plate
- Metal oxides – blackening, bluing, and phosphate treatments
- Coated with paints – organic paints, polyesters, acrylics, fluorocarbons
- Coated with glass – porcelain
- Oxide – weathering steel

A MATERIAL FOR ARTISTS

Steel and iron offer a significant, versatile metal for creating artistic forms and shapes. Methods of cutting, shaping, and joining are abundant and require only basic knowledge of metal working.

Perhaps the greatest concern for the artist is how the material will perform in the environment it will reside in, how the surface will be maintained, and what substances, particularly damaging substances, will find their way to the surface.

For steel artwork, even if it is coated, it is paramount to have a periodic maintenance program that removes potentially damaging materials from the surface. Painted surfaces are subject to fade and chalk as ultraviolet radiation unbinds the polymers and affects the pigments. Periodic maintenance cannot protect against this, but it can remove pollutants that accelerate the decay or create localized stains. Additionally, painted surfaces consistently have areas of porosity. These might be incredibly small but they exist on the surface. Edges and corners are common areas where the thickness of the paint film is less and prone to porosity in the surface. Surface tension characteristics and Faraday phenomena make it difficult to achieve a perfectly even surface coating on steel or iron.

Exposure to the atmosphere, to humid environments, and to moist urban environments expose the weakness in the film and can show rusting.

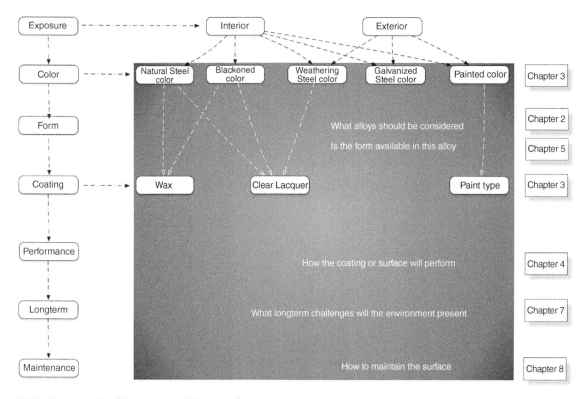

FIGURE 1.22   Steel in art and architecture decision map.
Source: L. William Zahner

A maintenance program removes corrosion particles that form at these porous regions, and it will assist in identifying areas where weakness exists so a record can be kept to determine when future remediation will be needed.

Steel has a high affinity to moisture, and any breach or susceptibility in the coating will attract and form an oxide. What results begins as a blister on the paint, usually along a weak edge. The paint appears to have lifted from the metal surface. The blister can grow laterally along the paint–metal interface until the paint flakes from the surface. The revealed metal surface will show the initial signs of exfoliation as the weak, spongy oxide develops on the exposed surface. Much of this was already formed under the blistered paint, causing the paint to locally delaminate from the metal.

CHOOSING THE RIGHT STEEL

From the designer's perspective, the choice of one material over another for a surface usually has roots in appearance. Performance, cost, availability, and how a surface will look over time in conjunction with other materials in the design all play a role in selection of one metal over another.

When steel is a material in consideration, this book can provide information of value to determine if steel is the correct choice. In a general sense, steel is available in one of five categories. Each category covers a broad range of color tones and surface textures. Figure 1.22 is a general map of information within this book. There are seven parameters down the left side that will influence the decision. The one parameter left off is *cost*. Cost weighs in at each level, depending on the decisions made.

The following chapters cover some of the territory of steel and hopefully shed light on what the various terms mean. Several of the alloys that are used in art and architecture are described with their chemical composition. It is recommended that you describe the nominal composition of the steel alloy you wish to use, as well as the mechanical properties and surface quality levels that the particular alloy will provide.

CHAPTER 2

Alloys

I can't see the forest because of the trees.

Proverb

INTRODUCTION

Steel has been around for centuries; however, its use as a natural surfacing material is only a somewhat recent architectural development, stainless steel being the exception. The scope of this book covers the carbon steels, which include the weathering steels. Stainless steels are discussed in length in another book in the series.

Note, to the extent of the steel alloys described, the focus is on those alloys that find use for their visual aesthetic. The vast majority of steel alloys are created for specific industries and have defined mechanical characteristics. Only a few of these will be presented here, the steels that find their way into the art and architectural world or that lie on the periphery of potential use.

The properties of steels have been adjusted and refined to achieve specific industry requirements far more than any other material, metal or otherwise, known to mankind. This is because for centuries, steel, its production, and use, have defined civilizations. In different parts of the world where civilizations flourished steel production played a key role. Today steel is used across many industries, but when we think of steel, we think of structures than are created to respond and resist loads. The breadth of steel usage as a structural component stretches across nearly every industry. Because of this, there are hundreds of customized steel alloys and multiple treatments to these alloys to create properties specific to the intended use.

In art and in architecture, steel is not normally considered as a surfacing material, except for the weathering steels. The weathering steels are desired for the natural color they display and their ability to withstand the onslaught of the environment. In art and architecture, for the past several decades, when steel has been considered, it has often been described as cold-rolled steel, Corten, carbon steel, wrought iron, cast iron, and several others including galvanized steel, which is more for the surface appearance than the core steel used.

The words used to describe steel are firmly rooted in the jargon of the industry that uses the metal and creates the raw material. Steel goes so far back in our human history with so many cultures and industrial endeavors, it is easy to see how terms used to describe the metal are a language in itself.

Terms such as *killed, semi-killed, rimmed,* and *capped* are used to describe types of carbon steel and relates to processes in manufacture. These terms have been replaced today by descriptive alpha numeric qualifiers such as CS for commercial steel, DDS for deep drawing steel, and further to EDDS or extra deep drawing steel.

The terms *cast iron* and *wrought iron* are reminiscent of the past. Today most wrought iron is in reality wrought steel. There are very few instances when pure iron is used today. Nearly all iron goes into the making of steels. Steels are an alloy of iron.

Steels contain less than 1% carbon and often have additions of silicon and manganese.

It is said that there are more than 3500 different steel alloys in use. Most of these alloys are created for specific industrial processes. For art and architecture, only a few alloys are generally considered. Aspects of strength, formability, welding characteristics, and surface finishing allow for a wide latitude, but the realities of availability and code limitations in structural applications narrow the available field considerably.

THE ALLOY NUMBERING CODE

In keeping with the Architectural Metal Series, the UNS system (Unified Numbering System) will be used as the primary designation for the alloys. Efforts will be made to cross reference industry-recognized alloy systems established by the major manufactures of steel. The UNS system uses an alpha-numeric system where the letter designates the family of the metal and the number is used to define a specific alloy within the family. See Table 2.1.

The most widely used system in North America is that created and revised over the years by the American Iron and Steel Institute (AISI) and the Society of Automotive Engineers (SAE). The AISI and the SAE International develop and maintain a standard numbering system for steel grades used extensively in North America. The numbering system has been in use for decades and is widely recognized within industry. The UNS numbering system has attempted to correlate with the AISI/SAE numbering system.

For example, in the AISI/SAE system, the carbon steels use a designation system that define the carbon content as well as the elements added. This is the North American method of defining the grade. The basic system uses four digits, sometimes with an additional alphanumeric value.

$$1 \, x \, x \, x$$

The first digit, 1 in this case, represents the steel as a carbon grade.

TABLE 2.1   Unified Numbering System designations.

UNS range	Type of metal	Comment
D00001 to D99999	Special steels	Not covered in this text
F00001 to F99999	Cast irons	Rarely in art and architecture
G00001 to G99999	Carbon and alloy steels	Most alloys in architecture are in this category
H00001 to H99999	Hardenability band; H steels	Not covered in this text
J00001 to J99999	Cast steels	Instead of cast iron, cast steel is used
K00001 to K99999	Misc. steels	Weathering steels
S00001 to S99999	Stainless steel	Covered in another text in the series
T00001 to T99999	Tool steels	Rarely in art and architecture

The second digit represents the added elements to the carbon steel. A designation of 1, represents sulfur while a 2 stands for sulfur and phosphate. These are processes used to alter the machinability of the steel. They are called *resulfurized,* the number 1, or *rephosphorized,* the number 2, in this instance.

The last two digits correspond to the approximate middle range of the carbon percentage. So, for example, 1018 would indicate 1 as the carbon grade, 0 for no elements added, and 18 for 0.15–0.20% carbon for approximate 0.18%.

The UNS categories incorporate these designations into their numbering system. In the above example, 1018 steel would be G10180. Seems straightforward so far; however, it can get more confusing as the logic of a particular industry designation is interpreted by other industries.

For example, when certain elements are added for a particular operation such as improved machining characteristics, the AISI/SAE system can add a letter in the middle. Take the case of AISI alloy 1214. Carbon alloy with approximately 0.14% carbon, with added phosphorus. Lead is added to make it a machining bar stock and the new designation is:

The equivalent UNS alloy designation is G12144.

Boron is another element sometimes added to the alloy, and it would be designated as a B occurring between the second and third numbers. An H added at the end of the number would designate a special hardening level.

Different industries often use other methods to define the steels that have been designed for a particular application. By design, these steels have specific elements or specific heat treatments included in their makeup.

TABLE 2.2   Comparisons of UNS G10180 and ASTM A36 steel.

Property	UNS G10180		ASTM A36	
Yield strength	240–400 MPa	35–58 ksi	250 MPa min.	36 ksi min.
Tensile strength	430–480 MPa	52–70 ksi	480 MPa min.	58 ksi min.
Elongation	17–27%		22%	
Hardness (Brinell)	130–140		140	
Machinability	Good		Fair	
Hot or cold rolled	Both (usually cold)		Hot rolled	
Surface finish	Good		Fair	
Chemistry	Tight controls		Larger range	

Common to the building construction industry is the steel designated as A36 steel. This is short for ASTM A36. ASTM, or ASTM International as it is called at this time,[1] is a recognized international standards organization involved with the development of standards for testing and materials.

When steel is designated A36, there are minimum mechanical properties that must be achieved when tested, since it is structural steel. A36 steel is similar to AISI 1018, or UNS G10180 steel with much tighter mechanical properties. UNS G10180 steel cannot be called A36 steel unless these minimum mechanical properties are achieved. These are wrought steels and are available in plate, sheet, rod, tube, and shapes. See Table 2.2 for comparisons.

In the Appendix C, cross references are made to the various alloy designation systems including the major European and Japanese designations.

ALLOYING ELEMENTS

Carbon

The main alloying element in steel and iron is carbon. It is odd to think that pure iron is soft, ductile, and never used in industry. The additions of very small amounts of carbon increase the strength and hardness enormously. Wrought iron, one of the first of the iron alloys created by early man, had just 0.05% carbon. This made the iron much stronger.

Today's steel has between 0.001% and 1.5% carbon. The variations created by just this small amount of carbon will have profound effects on the properties of steel. Cold working behavior, ductility, hardness, and heat treatments are all affected by this small amount of carbon.

Cast iron, another alloy form of iron, contains 2–4% carbon. This much carbon, however, makes cast iron brittle. It has excellent compressive strength and hardness, but it cannot take shock or

---

[1] ASTM was formerly known as the American Society for Testing and Materials. It was formed back in the late 1800s to establish standards for the steel used on railroad rails. There were more frequent rail brakeages that facilitated the need for an organization to establish minimum requirements for the manufacture of the steel.

bending because tiny carbon particles accumulate in the crystal structure as the iron carbon mix solidifies. Heat treatments will improve cast irons, making them more ductile but not to the same ductility of steels.

Other Elements

There are other elements added to steel and there are residual trace elements that find their way into the steel alloy from the various processes used to create the steel. The source of these trace elements today is usually recycled scrap; however, some residual elements can enter from the ore source itself.

Sulfur and phosphorus are controlled elements in steel production. In certain instances, small amounts are wanted, even added where in many other instances reducing the sulfur and phosphorus is an important step. Both these elements can have an appreciable effect on the properties and appearance of the steel. This is why most mill certifications will list the amount of sulfur and phosphorus in the steel.

C	Carbon	Carbon is the most important alloying element in steels. As carbon % increases, strength and hardness increases, but machinability and ductility decrease along with weldability. Most steels intended to be welded have less than 0.5% carbon.
S	Sulfur	Sulfur is added to aid in machining. Generally it is not a desired element found in steels and will affect the surface quality of the steel. Sulfur decreases weldability and toughness.
P	Phosphorus	Phosphorus is added in small quantities to increase strength and hardness as well as machinability. Like sulfur, it is not always desirable. It helps improve corrosion resistance in the low-alloy steels. Phosphorous reduces ductility and impact resistance.
Si	Silicon	Small amounts improve elasticity and toughness. Silicon is the principle deoxidizer. It promotes larger grain size. Increases the magnetic permeability. Added to cast steels to improve fluidity. In low-carbon steels it will affect surface quality.
B	Boron	Boron is added in small amounts to improve hardness.
Pb	Lead	Lead improves machinability but affects welding. The leaded carbon steels are considered free machining. Lead acts as an alloying lubricant in the machining operation with a reduction in friction and heat.
Mn	Manganese	Improves and stabilizes the heat treatment process. It is also a deoxidizer. Allows for a reduced quench rate, which gives more predictable mechanical properties. Manganese is a common alloying element found in steel. Manganese improves strength and hardness.
Cr	Chromium	In small amounts, chromium enhances the mechanical properties such as strength, hardness, and heat treatment ability.
Ni	Nickel	Improves hardness and impact resistance. Promotes austenitic grain formation.
Cu	Copper	Small amounts improve corrosion resistance.
V	Vanadium	Small additions help create a tighter, finer grain structure in steels. Vanadium improves toughness and creep resistance.
Te	Tellurium	Tellurium enhances the machining characteristics of carbon steels. Additions of small amounts of tellurium improve cutting and turning properties of the steel.
Se	Selenium	The element selenium is added in small amounts to improve machinability. Aids in the surface quality and appearance of the steel.
Al	Aluminum	Aluminum is added in small amounts as a deoxidizer of steel with or without silicon.

The two major classifications of metal are the cast and the wrought forms. Each form has specific characteristics unique to how they are created. All metals are first cast but, the wrought forms are refined further. These wrought forms are created by hot rolling into plate, extruded or drawn hot into shapes, forged and further rolled to shape, while some are further cold rolled into sheet, bar, tube, pipe, and wire. The wrought forms are widely used in the art and architectural industries. The cast forms were first used in architecture as beautiful storefronts and ornamentation but today the majority of uses come from the wrought classification.

TEMPERS

Often, the temper of low-carbon, cold-rolled steel is designated as a range defined by its hardness. There is a direct correlation between hardness and chemical makeup of the steel and between hardness and temper. When qualifying a temper desired, the chemistry range of the alloy should be stricter. For art and architecture, the tempers most commonly used can be found in Table 2.3.

THE WROUGHT ALLOYS OF IRON

The wrought alloys, like those of other metals, are the iron alloys that are rolled into sheet or plate, hot or cold drawn into bar or wire, hammered into plates, or shaped into tubes or pipe. The wrought alloy forms are distinguished from the cast forms even though they begin as a casting when the metal is first reduced from scrap or ore.

Early on, wrought iron was the commonly available form of iron. In many parts of the world, it actually proceeded cast iron. The energy needed to reach temperatures to melt iron in any appreciable amount simply could not be achieved. Iron-rich ores would be roasted in layers of charcoal, where air would be pumped in using bellows made from hide. Needles of iron and slag would be mixed with the ash of the roasted mixture. These small needles of iron would be hammered to form lumps intermixed with carbon-rich slag that accompanied the formation of needles. The result was a highly malleable, high-grade form of iron with a mixture of fibrous slag intermixed throughout.

TABLE 2.3   Tempers on cold-rolled steels.

Temper designation	Rockwell B hardness range
Quarter hard temper	B-60 to B-75
Half hard temper	B-70 to B-85
Full hard temper	B-85

WROUGHT IRON

Wrought iron was the major form until the mid-nineteenth century, when steel-making technology took shape and advanced the Industrial Revolution. Wrought iron is also called *puddled iron* due to the technique developed in the eighteenth century of melting the iron ore in a special furnace that would allow the formation of a puddle of refined iron for the production of wrought iron bars. The technique of stirring the molten iron was known in China as far back as the first century. Swords of superior strength could be made from this wrought iron. It was discovered that this form of iron could take a blow but still provide resilience and would not crack.

The more modern form of producing puddled iron developed in Great Britain in the late 1700s, and this advanced the industrial revolution by enabling larger amounts of iron bar without using charcoal as the fuel. Instead, coal was introduced as the source of energy.

The Eiffel Tower and the iron framework in the Stature of Liberty were all made from puddled iron. Steel was being manufactured at the time but was considered too new for the conservative engineering community. Eiffel himself wanted to use this new form of iron but ended up, after pressure from civic authorities, using the more common wrought iron instead. In a way, this was a good thing. True wrought iron has excellent corrosion resistance. The slag, glass-like stringers, in the wrought iron help prevent corrosion while providing strength. There are numerous instances of wrought iron fabrications over 100 years old still performing well. Figure 2.1 shows examples of wrought iron work over 100 years of age. The left image is wrought iron work in New Orleans. The right image is the work of Gaudi for the Palace of Guell. The lower left is a decorative wrought iron newel on a staircase in Europe.

Ferrite crystal structure is the main component within wrought iron. When wrought iron is heated above 900 °C, the body-centered cubic (BCC) structure of ferrite changes to a face-centered structure (FCC) of austenite. The FCC structure is more easily shaped by hammer blows than the BCC, so blacksmiths for centuries would heat the iron to a red-hot condition before hammering. The slag would liquify and act as a lubricant as the metal was hammered out. With wrought iron, welding sections together by heating until the metal is red hot is one unique attribute. Early iron plate was made this way. They were hammered by powerful weights moved by water wheels repeatedly hitting the metal and shaping it. Leonardo da Vinci devised a method of hammering metal into plates using the energy from a cam hammer invention driven by a water wheel.

In the early days of production, prior to the puddling process, small amounts of wrought iron would take 25 days to produce. One kilogram of wrought iron would consume as much as 25 kg of charcoal. The regions around these early producers of wrought iron were denuded of trees as the metal of war was in demand, and the environment suffered.

FIGURE 2.1   Images of wrought iron.

The process to make wrought iron changed later in the eighteenth century. The Aston, or the Aston Byers process as it is sometimes referred, was developed. This process starts with pig iron made from the Bessemer converter. The pig iron is melted and poured into molten slag and mixed together. A spongy mass is produced, and the wrought iron is rolled out. This increased the production rates necessary to provide adequate amounts of wrought iron. Today, true wrought iron is rare. It is produced mainly by recycling, dismantled old wrought iron or ship mooring chains. What we call wrought iron is actually low-carbon steel. Figure 2.2 shows conventional forms made to look like the wrought iron of old.

FIGURE 2.2   Steel rails and chandeliers made to look like wrought iron of old.

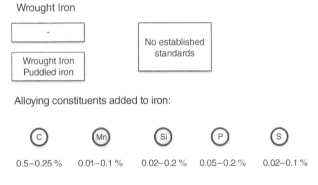

Wrought Iron

-

| Wrought Iron |
| Puddled iron |

| No established |
| standards |

Alloying constituents added to iron:

C	Mn	Si	P	S
0.5–0.25 %	0.01–0.1 %	0.02–0.2 %	0.05–0.2 %	0.02–0.1 %

	Tensile strength range		Yield strength range	
Wrought iron	34–54 ksi	234–372 MPa	23–32 ksi	159–221 MPa

Alloying constituents and mechanical properties of the early wrought iron were highly variable. Wrought iron can be welded but the slag that is intermixed with the iron can interfere with accomplishing a good weld. The low carbon content prevents wrought iron from being able to be heat treated to control mechanical properties more effectively.

Wrought iron soon gave way to wrought steel. The mechanical properties of steel, the ability to make large amounts of steel, and the new energy source of electricity all contributed to the rise of steel. Today, what we see and refer to as "wrought iron" is actually mild steel. As shown in Figure 2.2, "Wrought iron" handrails, chandeliers, and other work made from steel requires constant maintenance to survive against the forces that lead to corrosion.

THE WROUGHT ALLOYS OF STEEL

There are several basic forms of wrought steel. These are the carbon steels, alloy steels, high-strength, low-alloy (HSLA) carbon steels, tool steels, and what are designated as ASTM steels.

Within the carbon steels, there are the low-, medium-, and high-carbon steel designations. These are defined by the range of carbon added to the iron. There are other elements introduced as well. For the purpose of surfaces used in art and architecture, these other elements make up trace amounts and usually are not of consequence. Table 2.4 shows the three classifications of the carbon steels and the carbon content.

Of these carbon steels, there are only a few in common use for surfacing materials in art and architecture. Reference Table 2.5. Most fall under the low carbon steel classification and are

TABLE 2.4 Classifications of carbon steels.

Low-carbon steel	0.06–0.25% carbon	Mild steel	G10050–G10260
Medium-carbon steel	0.25–0.55% carbon	Medium steel	G10290–G10530
High-carbon steel	0.55–1.00% carbon	Hard steel	G10550–G10950

TABLE 2.5 Low-carbon steels used in art and architecture.

Category	UNS No.	AISI/SAE ASTM
Carbon steel	G10060	1006
Carbon steel	G10080	1008
Carbon steel	G10100	1010
Carbon steel	G10180	1018
Carbon steel		ASTM A36

TABLE 2.6 Physical and mechanical properties of steels.

Physical properties	Mechanical properties
Thermal expansion	Stress and stain
Electrical conductivity	Elasticity – both elastic and plastic
Color	Tensile strength
Texture	Ductility
	Hardness
	Impact resistance
	Fatigue resistance

determined by less stringent qualifications than other industries require. The classification A36 is also in this carbon steel class. Note the A36 nomenclature corresponds to stringent mechanical properties more so than chemical composition and makeup of the steel.

With the steels, physical properties do not change as much as the mechanical properties with changes in the microstructure of steel as carbon is added. Physical properties are influenced by the interatomic forces at the submicroscopic level while mechanical properties are influenced by changes in the lattice structure. Reference Table 2.6.

Thus, in the context of appearance, steels surfaces are not dependent on the makeup of the alloy. The surface texture and to a point, the color and appearance are determined more by the hot or cold rolling of the final surface. The cold-rolled surface of the various steel alloys appear similar, as do the hot-rolled surfaces, regardless of the alloy.

STEEL ALLOY G10060

UNS G10060

| ASTM A510 |
| ASTM A29 |
| ASTM A545 |

| 1006 |
| Carbon steel |

Alloying constituents added to iron:

C	Mn	P	S
0.08% max.	0.25–0.40%	0.40% max.	0.05% max.

	Tensile strength		Yield strength		Elongation % (2 in.; 50 mm)	Rockwell B hardness
G10060	48 ksi	330 MPa	41 ksi	285 MPa	20	95

Available Forms

Sheet – cold-rolled

Plate

Extrusion

Tube

Wire

Alloy G10060 is ductile and soft. This is an alloy used in industries that require significant forming and shaping such as the automotive and appliance industry. The mechanical properties are dependent on temper.

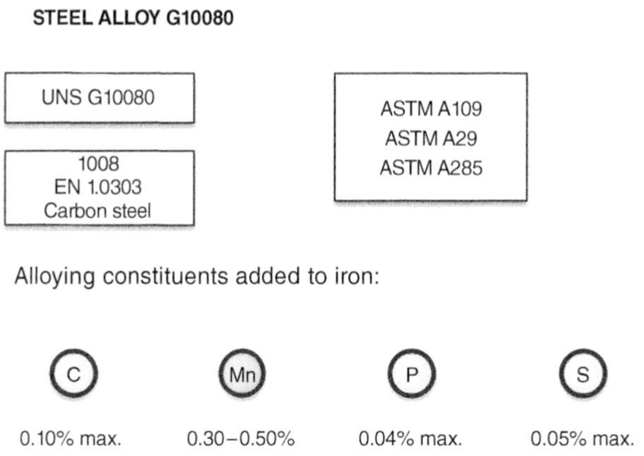

STEEL ALLOY G10080

UNS G10080

1008
EN 1.0303
Carbon steel

ASTM A109
ASTM A29
ASTM A285

Alloying constituents added to iron:

C	Mn	P	S
0.10% max.	0.30–0.50%	0.04% max.	0.05% max.

	Tensile strength		Yield strength		Elongation % (2 in.; 50 mm)	Rockwell B hardness
G10080	49 ksi	340 MPa	41 ksi	285 MPa	20	55

Available Forms

Sheet – cold- and hot-rolled

Plate – hot-rolled

Extrusion

Tube

Wire

UNS alloy G10080 is very similar to alloy G10100. Often, they are quoted as an either/or, depending on availability. This alloy is produced cold-rolled sheet steel. It is available in all quality levels, commercial steel or commercial quality, both hot and cold, drawing quality and in structural quality. When ordering simply mild steel or carbon steel sheet, this is the common alloy provided by many service centers and supply houses. Figure 2.3 shows a typical G10080 surface that has been blackened.

Used where fabricated products are produced from significant forming operations, this alloy has good weldability and has a high thermal conductivity, compared to other steels. Alloy G10080 is not as strong as G10100, but you gain some improvement in formability. There is a wide range of strength characteristics that depend on how the steel has been tempered.

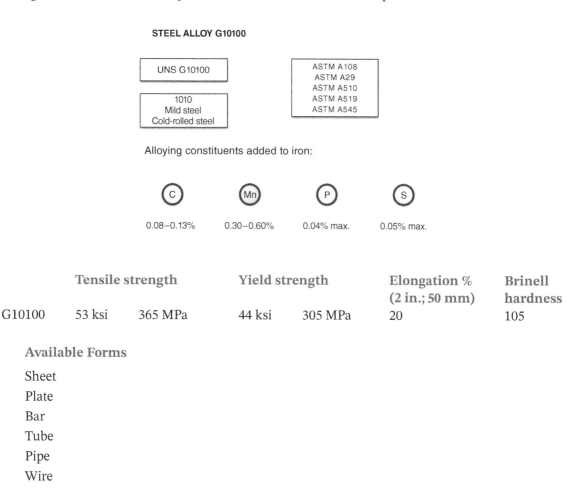

STEEL ALLOY G10100

UNS G10100		ASTM A108
		ASTM A29
		ASTM A510
1010		ASTM A519
Mild steel		ASTM A545
Cold-rolled steel		

Alloying constituents added to iron:

C	Mn	P	S
0.08–0.13%	0.30–0.60%	0.04% max.	0.05% max.

	Tensile strength		Yield strength		Elongation % (2 in.; 50 mm)	Brinell hardness
G10100	53 ksi	365 MPa	44 ksi	305 MPa	20	105

Available Forms

Sheet

Plate

Bar

Tube

Pipe

Wire

Alloy G10100 is commonly available as low-carbon, mild steel. Many auto bodies, for example, are made from this alloy of steel. It has good ductility and weldability. It is available in the annealed

FIGURE 2.3   Blackened G10080 steel surface used on interior surfaces.

state, but the tensile strength can be markedly increased by tempering. Ultimate strength can be increased to as much as 75 ksi (517 MPa) when quenched and tempered. This alloy is typically more economical than alloy G10080 because the composition is less exacting.

A 36 Steel

At this point it is worth describing what is meant by A36 steel. A36 is a common structural steel, with a maximum 0.3% carbon. A36 is short for ASTM – A36, as defined by ASTM International, the former American Society of Testing and Materials. For a steel to be classified as A36 steel, it must possess certain mechanical attributes. It is defined more by mechanical characteristics than by the alloying constituents.

This steel type is very widely used, generally for structural purposes.

The UNS number for this steel type is K02600.

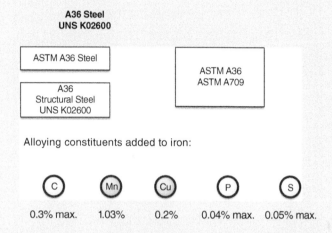

Mechanical Properties of A36 Steel

Tensile strength:	58–80 ksi	400–550 MPa
Yield strength:	**36 ksi minimum**	**250 MPa minimum**
Elongation	20%	
Modulus of elasticity	29 ksi	200 MPa
Shear modulus	12 ksi	79 MPa

Available Forms

Hot-rolled plates

Hot-rolled bars

Structural shapes

A36 steels have a maximum 0.3% carbon, which limits the effect of heat treatment. These steels have good weldability and are nearly always provided in hot-rolled forms of plates and bars. Figure 2.4 shows an artform made of A36 steel.

(*continued*)

(continued)

FIGURE 2.4   A36 steel, welded, shaped, and painted to create art form.

These steels have very good impact resistance. When A36 steel is specified, it must be tested for mechanical behavior. It is very important to understand that you cannot simply use any carbon steel when the A36 requirement is presented.

These are SQ, *structural quality* steels.

ASTM A570 defines the standard for structural quality hot-rolled sheet and plate. ASTM A611 defines the standard for structural quality cold-rolled sheet.

STEEL ALLOY G10180

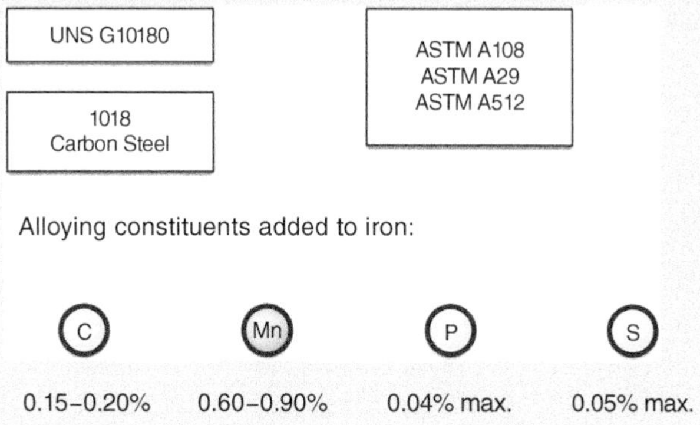

UNS G10180		ASTM A108
		ASTM A29
1018		ASTM A512
Carbon Steel		

Alloying constituents added to iron:

C	Mn	P	S
0.15–0.20%	0.60–0.90%	0.04% max.	0.05% max.

	Tensile strength		Yield strength		Elongation % (2 in.; 50 mm)	Brinell hardness
G10180	63 ksi	440 MPa	54 ksi	370 MPa	15	126

Available Forms

Cold-formed bars

Cold-drawn shapes

As noted, this alloy has a broader range of allowable mechanical properties than A36 steels. Refer to Table 2.2. This is due to the higher amount of carbon and manganese. Note that the hardness of this alloy is significantly greater than other carbon steels. This alloy has good formability and can be case hardened.

This alloy is considered for machining and turning operations such as shafts, screw guides, and other machined forms. Alloy G10180 is referred to as a free machining grade of steel. This alloy has good welding characteristics and has a good surface finish.

There are other carbon steel alloys that are rarely used in art and architecture. Alloy G10200, G10250, G11170, G10440, and G10450 are a few that make up bolts and fasteners, special hardened rods and bars, and forged and machined parts. They are not common alloys in the general architectural or ornamental surfacing field.

STEEL ALLOY G10200

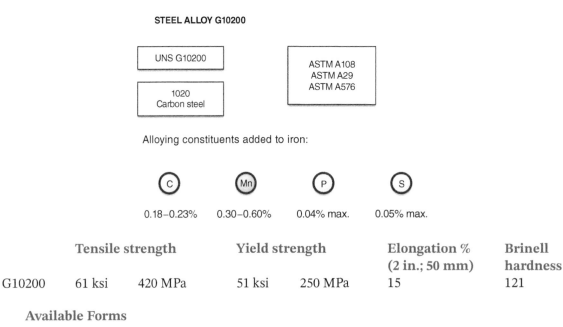

UNS G10200

1020
Carbon steel

ASTM A108
ASTM A29
ASTM A576

Alloying constituents added to iron:

C   Mn   P   S

0.18–0.23%   0.30–0.60%   0.04% max.   0.05% max.

	Tensile strength		Yield strength		Elongation % (2 in.; 50 mm)	Brinell hardness
G10200	61 ksi	420 MPa	51 ksi	250 MPa	15	121

Available Forms
Cold-formed bars
Cold-drawn shapes
Hot-rolled bar

This alloy can be hot and cold worked. Alloy G10200 is the alloy used in pins, bolts, chains, and other service parts that need good workability and strength.

THE WEATHERING STEEL ALLOYS – HIGH-STRENGTH, LOW-ALLOY STEELS

The weathering steels are a special alloy steel used extensively in art, architecture, exposed structures, and landscape architectural elements. There are a number of high-strength, low-alloy (HSLA) steels, but only a few find use in art and architecture.

Often referred to by the trade name, COR-TEN®, an alloy developed by US Steel back in the 1930s, these steel alloys are categorized by grades. The grades represent different but specific alloying ranges. However, for the purposes of art and architecture, and the expected exposures of these steels, the grades A and B are used.

What makes these weathering steels unique and special is the tight, adherent oxide that develops on the surface. When these steel surfaces are exposed to moisture, they go through similar surface oxide formations as carbon steels but, as the surface corrosion continues, the alloying elements create a dense, adherent hydroxide that slows the corrosion rate way down as compared to the loose, fragmented oxide that develops on typical, low-carbon steels. The actual mechanism of this protective process is covered in more detail further on in this book.

In art and architecture, it is the color and texture this oxide exhibits when it has fully developed that is of interest as well as the corrosion resistance behavior. At first the thought of a rusty old surface may be paradoxical. In the modern sense, usually you want a surface not to rust. Rusting metal is synonymous with decay so why would one want a surface that is decaying? Initial stages may appear this way and cause a bit of angst but with exposure and time the surface takes on a very natural, sometimes rugged, beauty. See Figures 2.5 and 2.6.

FIGURE 2.5   Weathering Steel "sound wall" Stanford University.
Source: Designed by Legorreta Architecture and Frida Excobedo.

FIGURE 2.6   Detail view of weathering steel at Stanford.
Source: Designed by Legorreta Architecture and Frida Excobedo.

Today, many projects use preweathered, weathering steel. This way when the surface created with the metal is presented it has already developed much of the color and tone that takes months to achieve. Preweathering is performed by custom metal finishing facilities that are expert in the surfacing of steel and exercise environmental prowess to avoid harm to the environment or personnel working with the metal.

Weathering steels have very good strength and can be welded. They are ductile but hard. After years of exposure, the weathering surface becomes very hard and will resist abrasion. Initially, however, the surface is crystalline and can be abraded easily.

The UNS categorizes these steels in the miscellaneous category with the alphanumeric code beginning with a K. Most specifications in North America, qualify these alloys by the ASTM number corresponding to the form.

For plates 5–50 mm thick	ASTM 588 – A
For shapes, angles, and channels	ASTM 588 – B
For sheets 2–10 mm thick	ASTM 606 – 4
Structural steel shapes	ASTM A709 – 50W
Welded steel rectangular pipe, tube	ASTM A847

The ASTM Standard A588 has the added letter to refer to the grade. For what is used in art and architecture, the two grades, A and B, are listed. There are additional grades corresponding to different chemical makeup. For example, Grade A contains carbon, manganese, chromium copper, and nickel to specific ranges. Grade J contains carbon and manganese only as the alloying constituents.

STEEL ALLOY K11430

UNS K11430		ASTM A588
		ASTM A606
A588 Grade A		ASTM A242
Corten		
EN 1.8959		
S355JOW		

Alloying constituents added to iron:

C	Mn	Cr	Si	Cu	Ni
0–0.19%	0.8–1.5%	0.4–0.65%	0.3–0.65%	0.25–0.4%	0.4% max.

P	S	V
0.04% max.	0.05% max.	0.02–0.1%

	Tensile strength		Yield strength		Elongation % (2 in.; 50 mm)	Brinell hardness
K11430	70 ksi	485 MPa	50 ksi	345 MPa	16	170

Available Forms
Sheet – Covered under ASTM 606-4
Plate

Bar

Tube – Covered under ASTM A847

Pipe – Covered under ASTM A847

STEEL ALLOY K12043

UNS K12043

ASTM A588 ASTM A606 ASTM A242

A588 Grade B Corten B S355J2P

Alloying constituents added to iron:

C	Mn	Cr	Si	Cu	Ni
0–0.2%	0.75–1.5%	0.4–0.7%	0.15–0.5%	0.2–0.4%	0.5% max.

P	S	V
0.04% max.	0.05% max.	0.01–0.1%

	Tensile strength		Yield strength		Elongation % (2 in.; 50 mm)	Brinell hardness
K12043	70 ksi	485 MPa	50 ksi	345 MPa	16	170

Available Forms

Sheet – Covered under ASTM 606-4

Plate

Bar

Tube – Covered under ASTM A847

Pipe – Covered under ASTM A847

These are the weathering steel alloys that are commonly used in art and architecture. These alloys have excellent strength. The yield strength is a minimum 345 MPa (50 ksi) which is significantly higher that A36 steel. These steels have exceptional hardness. As the oxide grows and thickens on the surface of the weathering steels, their corrosion resistance in normal atmospheres will extend

FIGURE 2.7   Preweathered K11430 weathering steel used on the interior and exterior of the Communications School in Doha, Qatar.
Source: Designed by Antoine Predock.

the service life of these metals for decades. Figure 2.7 shows a preweathered steel surface used both interior and exterior of a building in Doha, Qatar.

It is important to note that slight variations in alloying constituents from European and Asian producers are expected. However, the mechanical properties should be similar. Many of these alloys are used in structural applications, and it is important to ensure the mechanical properties are sufficient to accommodate the design loads and ductile requirements.

OTHER STEEL ALLOYS

The Alloy Steels

Alloy steels are not used in art and architecture as a surfacing material. This designation covers the steels that possess higher levels of manganese, silicon, or copper than the other carbon steels. Some of these alloys contain nickel and chromium as well as molybdenum but lower levels than what would define them as stainless steels. The alloy steels are used in specialized industrial applications such as bearings, fasteners and bolts, axles, pressure vessels, saw blades, and aerospace (Table 2.7).

Alloy G41300 is listed because it is one of the more stocked steels due to its improved corrosion resistance and machinability. This alloy is used in forging operations. It has good forging characteristics when hot forged.

STEEL ALLOY G41300

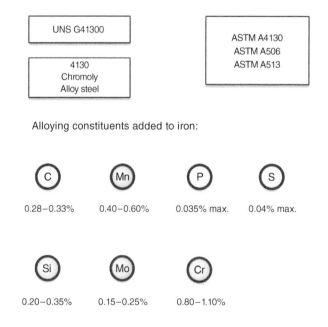

TABLE 2.7   Alloy steels.

UNS number
G13300–G13450
G40230–G48200
G51170–G52986
G61180–G61500
G86150–G92600

	Tensile strength		Yield strength		Elongation % (2 in.; 50 mm)	Brinell hardness
G41300	81 ksi	560 MPa	67 ksi	460 MPa	21.5	217

Available Forms
Sheet
Plate
Bar
Tube
Pipe

Alloy G41300 is used in aerospace, oil and gas and other operations where good machinability, strength, and corrosion resistance is critical. The chromium and molybdenum add strength and hardness as well as improved corrosion resistance.

This alloy is not commonly used in art or architecture.

Tool Steels

This steel category is rarely used in art or architecture. Tool steels is a classification of steel that has a higher carbon content than other steels and often has various alloying elements added to induce specific characteristics such as impact and shock resistance, cold or hot working parameters and other specific fabrication qualities. Alloying elements such as tungsten and vanadium are added as well as small amounts of chromium and nickel. On occasion they are found in components that have been manufactured in machine shops where the steel is more commonly used.

Tool steels possess tighter parameters from the perspective of alloying components and mechanical properties. These steels undergo a significant amount of quality control throughout their development and application. The grades of tool steels are subjected to more exacting requirements than structural steel grades. However, they should not be substituted for structural steels, unless they have undergone testing and verification to meet the requirements established for structural steel components. These steels have more carbon in them which improves their hardness and wear resistance beyond that of the mild steels.

Within the tool steels there are several categories as defined by the AISI. There are tool steels with a hardened outer surface produced by quenching in water to case harden the outer layer. These are called W steels for *water hardened*. These steels have higher levels of carbon, 0.6% to as much as 1.15%.

There are also tool steels designated as *shock-resistant*. These steels are designed as S steels. They have excellent toughness when it comes to shock loading. Some of these alloys exhibit the ability to absorb high levels of energy before fracture. The shock-resistant tool steels are carbon steels with additions of silicon and chromium.

Other tool steels fall under one of the categories of cold-work tool steels. These are the "oil-hardened" tool steels and "air-hardened" tool steels designated by the letter O or A. These tool steels are used in applications where dimensional accuracy and stability is required. They have good toughness and hardness.

Other tool steels designed for specific applications are the high chromium tool steels designated by the letter D. These are similar to some of the stainless steel alloys. There are the mold steels, designated by the letter P. These are low-carbon alloys, some with as little as 0.10% carbon and are designed for making the molds for injection plastic molding operations.

There are the *hot work tool steels*, designated by the letter H. These alloys contain high levels of chromium. The hot work tool steels are used to make dies for casting aluminum and zinc die cast products and for manufacture of punches used for piercing.

More specialized steels in the tool steel category are the T steels or high-speed steels. These contain tungsten as the principal alloying element. These special steel alloys are used to create high-speed cutting and drilling tools. They have extremely high abrasion resistance.

In addition to the T steels, there are tool steels called *maraging steels*. Here nickel is the primary alloying element. They have a very low carbon content of 0.03% maximum. These are very-high-strength steels used in the aerospace industry, as well as die casting and applications where holding sharpness is critical. The UNS number for tool steels begins with T followed by five numeric values for the specific alloy.

A few examples of tool steels are described in Table 2.8.

TABLE 2.8   Examples of tool steels.

UNS	Carbon (%)	Other elements	Applications
T11201	0.85	Cr, Ni, Mo, W, V	Drills, saw blades, lathe tooling
T30102	1.00	Cr, Ni, Mo, V	Punch tooling and embossing tooling
T30402	1.50	Cr, Ni, Mo	Steel blades and cutlery

CAST ALLOYS

Casting of iron into useful tools and shapes dates back several millennia. It took until the time when sufficient energy sources could generate the temperatures needed to melt iron. It was not until the seventeenth century in Europe with the development of the crucible process, that casting of iron was possible, whereas in ancient China, iron casting was an art practiced since the fifth century BCE.[2] The Chinese discovered the means of developing the high temperatures needed to melt iron and how adding carbon dropped the temperature needed to melt the metal.

Cast iron was in extensive use across Europe and the United States for much of the nineteenth century and early twentieth century. Many of the castings were single units such as the cast door in the left image of Figure 2.8 or assembled from several casting, such as the cast iron fountain in Washington, DC, on the right.

Cast iron comes in a number of alloys, generally falling into one of these categories:

- Gray cast iron
- White cast iron
- Malleable cast iron
- Ductile cast iron

Of these, the gray cast iron alloys are more common today as castings for art and architecture. The other cast alloys are for specialized industries such as automotive and tooling.

White cast iron is brittle. It contains carbide impurities that allow a crack to propagate unstopped through the casting. It is a very hard alloy and finds uses where extreme hardness is needed.

Malleable cast iron and ductile cast iron are used in automotive and other applications where special tempering is required for a particular product. For example, the malleable cast irons, UNS F20000 to F20005 are used in automotive engine blocks. These alloys have low ductility but are resistant to elevated temperatures.

The ductile cast irons, also called *nodular* due to spherical nodules of graphite are quenched and tempered to various predesigned levels. The UNS alloy numbers for a few of these are F22800, F33100, F33800, F34100, F34800, F36200.

Gray Cast Iron

This is the most common cast form of iron used in art and architecture. Gray cast iron gets its name from the appearance of the inside core of the metal when fractured. The color is a granular gray color

---

[2]Young, S.M., Pollard, A.M., et al. (1999). *Metals in Antiquity*, 1–9. Oxford: Archaeopress.

FIGURE 2.8   Cast iron door and cast iron fountain.

due to the formation of graphite within the casting. The graphite has a platelet or flaky appearance. This improves the strength and inhibits fracture propagation when cracks are initiated.

Figure 2.9 shows gray cast iron used to create roof shingles on a subway entrance in New York and as compressive support plates lining the interior of a subway in London.

There are approximately 11 common variations of gray cast iron designated by various grades established by ASTM A48. There is a wide range of properties exhibited by these grades depending on heat treatment and alloying levels. See Table 2.9.

When using these cast iron alloys, work with the foundry to determine the best strength for the particular application. All the gray cast irons fall in the ranges of alloying constituents, as shown below.

Grey Cast Iron

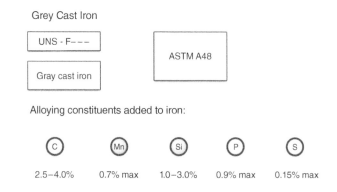

UNS - F- - - -

ASTM A48

Gray cast iron

Alloying constituents added to iron:

C	Mn	Si	P	S
2.5–4.0%	0.7% max	1.0–3.0%	0.9% max	0.15% max

FIGURE 2.9 Cast iron used as roofing tiles and as an interior liner for a subway tunnel.

TABLE 2.9   Various alloys and the associated grades of several gray cast irons.

Alloy	ASTM	Characteristic
F11401	Grade 20	As fabricated temper, low strength
F11701	Grade 25	As fabricated temper, low strength
F12101	Grade 30	As fabricated temper, average strength
F12401	Grade 35	As fabricated temper, average strength
F12801	Grade 40	As fabricated temper, average strength
F13101	Grade 45	As fabricated temper, moderate strength
F13501	Grade 50	As fabricated temper, moderate strength
F13801	Grade 55	As fabricated temper, high strength
F14101	Grade 60	As fabricated temper, high strength

	Tensile strength range of alloys		Yield strength range of alloys		Elongation % (2 in.; 50 mm)	Brinell hardness
Gray cast irons	23–65 ksi	160–450 MPa	14–43 ksi	98–290 MPa	0.5–9.6	160–300

The compressive strength of the gray cast irons is important because normally these forms should not be used where tension is a significant applied force. Subjecting these to bending stress should be avoided as well. Compressive strength of the gray cast irons is 83–190 ksi or 570–1290 MPa.

Examples of decorative gray cast iron railing in New Orleans are shown in Figure 2.10. This is a testament to the strength, beauty, and corrosion resistance of gray cast iron.

White Cast Iron

White cast iron is not normally used in art and architecture today.

White cast iron gets its name from the appearance of the inside of the casting when it fractures. Carbon forms in the iron solid solution as carbides. The white color is from a substance called cementite, iron carbide, that make this cast alloy much harder but brittle. When the molten metal of the cast iron cools rapidly and the carbon is not able to form graphite, if forms cementite and pearlite, whitish granular minerals in the iron solid solution. White cast iron lacks the ability to absorb energy from impact and when it cracks, there are no graphite flakes to stop the crack from going through the casting.

White cast iron is very hard and abrasion resistant. Not machinable and should be cast to the final shape, its hardness will thwart any additional surface work.

FIGURE 2.10   Gray cast iron railings of the LeBranche House, New Orleans.

White cast iron has very high compressive strength and the Brinell hardness can be as great as 600, twice that of gray cast iron.

White cast iron was used extensively in ancient China, and there are white cast iron artifacts that date back to the tenth century. The size, nature of the casting, and the corrosion resistance are remarkable on many of these objects. The intricate iron pagoda of the Guangxiao Temple build in 967 CE is assembled from several stacked cast iron elements.

Another example demonstrating the skill and abilities of the early Chinese metal workers in casting iron is shown in Figure 2.11. The giant casting of a lion, known as the "Sea Guard Howler" is located in Cangzhou City, China. This is a tenth century casting made of several sections assembled together. It stands 5.4 m (18 ft) tall and weighs approximately 40 000 kg (88 200 lb).

Malleable Cast Iron

Malleable cast iron has a high melting point. It is used extensively in engine blocks and other applications where high temperature resistance is needed. Malleable cast iron lacks ductileness. It is made from heat treatment of white cast iron. There are three types of malleable cast irons.

FIGURE 2.11   Giant lion casting of Cangzhou City. Cast from white cast iron.

- Whiteheart – contains 2.8–3.4% carbon
- Blackheart – contains 2.2–2.8% carbon
- Pearlitic – where the graphite has mostly changed to pearlite

The malleable cast irons are being replaced by the ductile cast irons because of the higher cost and time required to convert white cast iron to malleable cast iron. These are forms developed centuries ago. Whiteheart is a European version of this cast iron and was developed in 1722. Blackheart was developed in America in 1820.

The more common of the malleable cast irons are in Table 2.10.

TABLE 2.10 Malleable cast irons.

Alloy	Carbon (%)	Silicon (%)	Manganese
F22200	2.3–2.7	1.00–1.75	0.55% max
F23131	2.4–2.7	1.25–1.55	0.55% max

Ductile Cast Iron

Ductile cast iron has good strength and toughness, superior to that of gray iron. Also known as nodular cast iron due to the formation of graphite nodules instead of the flakes seen in gray cast iron. Developed in the 1940s, this hard, yet ductile alloy has magnesium added to cause the graphite to form into the nodules or spheres within the iron.

Ductile cast iron has application across industries where strength and ductility are needed. Mechanical properties are more critical than alloying composition for the ductile cast irons.

For example, the alloy F32800 is also known in industry as 60-40-18 ductile cast iron. This stands for:

Tensile strength of 60 ksi minimum (414 MPa)

Yield strength of 40 ksi minimum (276 MPa)

Elongation of 18%

Other alloys can have strength requirements double these values. But it is the mechanical properties that define these alloys for specific industry needs.

The ductile cast irons contain 3.50–3.80% carbon, around 0.05% magnesium to create the nodular formation of the graphite. They contain small amounts of manganese, less than 0.05%, 1.7% to close to 3% silicon, traces of nickel, phosphorus, and even copper.

CAST STEELS

Cast steels are used in various industries where complex shapes and metallurgic constraints are the controlling factors. Cast steels differ in several ways from the cast irons. From an art and architectural standpoint, ease of casting and economy of the cast irons are often the deciding factor. A few of the major differences between the two materials are listed in Table 2.11.

For many reasons, the cast steels are destined for industrial applications rather than architectural applications. Cast iron is often used for art castings and architectural applications where mechanical behavior is less a concern.

TABLE 2.11   Comparisons of cast steel to cast irons.

Characteristic	Cast steel	Cast iron
Carbon content	<2.0%	>2.0%
Corrosion resistance	Low	Moderate
Cost	Moderate	Low
Shrinkage	High	Low
Energy needed	Moderate	Low
Castability	Difficult	Easy
Flowability	Moderate	Good
Vibration dampening	Poor	Good
Wear resistance	Poor	Good
Tensile strength	Good	Poor
Elongation	Good	Poor
Ductility	Moderate	Good
Difficulty in production	Moderate	More defects
Design flexibility	Excellent	Good
Complex shapes	Excellent	Fair
Metallurgic flexibility	Good controls	Not as flexible
Weldability	Good	Poor

Both cast irons and cast steels can be galvanized in the hot dip process to improve corrosion resistance.

CHAPTER 3

Finishes

I am of the nature to grow old. There is no way to escape growing old.
One of the Buddhist 5 Remembrances.

Thich Nhat Hanh

Source: The Heart of the Buddha's Teaching: Transforming Suffering into Peace, Joy, and Liberation by Thich Nhat Hanh. ©1998 Broadway Books.

INTRODUCTION

When we think of the appearance of steel, we envision the rough texture of a cast surface or the semi-smooth gray blue of a structural beam. Weathering steel and the reddish brown layer of deep oxide may also come to mind, when we think of art or architectural uses of steel. For those that work with the metal from the context of an aesthetic surface, steel can have many possibilities.

In the right hands the appearance of a steel surface can be strikingly beautiful. Deep blacks, blues to variable black grays, olive greens, and other tones are possible. Figure 3.1 shows a blackened steel wall surface.

Preweathering the high-strength, low-alloy (HSLA) steels, commonly referred to as Corten, can produce deep purplish brown tones or vibrant orange red. Variations across the surface can be introduced and expanded upon as this amazing metal reacts and changes. The surface created is a result of the natural consequences of the formation of various iron oxides that seal the surface from further corrosion. Figure 3.2 is an example of the intense beauty weathering steels can display.

Steel is considered for its strength and hardness however, its natural mill surface appearance or the chemically enhanced surface can add a natural beauty that is as engaging as other natural surfaces such as wood, copper, and stone.

FIGURE 3.1   Blackened steel with butterflies.
Source: Courtesy of Imagewall™.

FIGURE 3.2   Weathering steel used on the New Science Center; Amherst College designed by Payette.
Source: Courtesy of Jeffrey Abramson.

When considering steel as an aesthetic surface material it is crucial to understand the differences that exist in production processes and the way these will determine the appearance of the steel. Note this is for the wrought forms of sheet and plate material. Cast steel surfaces reveal a basic textural appearance generated from the casting method used whereas sheet and plate have the surface defined by the process of manufacture and the subsequent treatments used. For the most part though, raw, untreated steel sheet or plate will appear very similar. Like the wood from an oak tree, they are of the same "species," just not exactly the same.

THE MILL SURFACE OF SHEET AND PLATE MATERIAL

Hot rolled or cold rolled. Hot drawn or cold drawn. These produce the surfaces for steel wrought products more so than the alloy variations. Even for the weathering steels with their deep oxide development on the surface, the differences in hot and cold rolled will induce surface variations.

The difference resides in the surface quality and character as well as the color. Hot-rolled steels tend to be darker in color, have a rougher surface and are thicker. Cold-rolled steels are smoother, have finer grains and the color is lighter. Figure 3.3 shows an interior surface made from the natural steel appearance.

The Carbon Steel Surface

The carbon steels come in two major forms, hot rolled and cold rolled.

FIGURE 3.3   Natural steel surface used on the interior of the Oakley Headquarters.
Source: L. William Zahner.

On the low-carbon steels, the hot-rolled surface is darker and slightly rougher than the cold-rolled surface. It can be given a skim pass through cold rolls to improve the clarity and smoothness of the hot-rolled surface.

Hot-Rolled Surface

Most steels today is produced by the continuous casting method, Chapter 1 shows the basic aspects of continuous casting. As the metal is cast and rolled hot in a large, continuous slab, the grains of the metal are extended in the direction of rolling. The steel is kept hot, approximately 926 °C (1700 °F), which is above the temperature where the steel recrystallizes. Hot rolling at these temperatures allows for easier shaping and forming. Structural steel (SS) shapes are formed at or above, these temperatures.

As the steel slab continues to be rolled hot, the steel is stretched and reduced in thickness. As the steel cools it will shrink, so dimensional controls of thickness are not precise. You can expect slight variations all within the allowable tolerances.

The surface develops scale, a tough, glass-like oxide. This oxide form on the hot surfaces, creating visual surface abnormalities. Mill scale thickens and elongates on the surface as the hot-rolling process continues. Figure 3.4 shows a thick steel plate with scale.

Scale is a combination of three oxides of iron, FeO, Fe_3O_4, and Fe_2O_3. It will protect the underlying steel from atmospheric corrosion attack unless it has a crack or break in it. However, scale is cathodic to the base, surrounding steel and if broken will cause the steel to oxidize faster. The oxide on thick steel plates is thicker and less adherent. It can flake off or be removed more readily by abrasive blasting, but an impression is left behind on the surface. Scale is blue-black in appearance and lacks the flexibility of the steel around it.

FIGURE 3.4   Scale on 50 mm thick steel plates.
Source: L. William Zahner.

For most steel products, the scale is removed after hot rolling. Scale will interfere with coatings, can damage dies on subsequent cold processing and reduces the corrosion resistance. Scale is removed by abrasive blasting or a process called pickling. For most hot-rolled sheet and plate, the scale is removed by pickling. Pickling involves powerful organic acids, usually hydrochloric or sulfuric acids applied either continuously or by immersing the plates into the acids.

On thinner sheets the pickling will remove small scale particles and the surface may show blemishes or small pits. Hot-rolled sheets can be further rolled in tempering rolls and flattening rolls as well to improve the surface appearance.

After the hot rolling, the steel is considered to have a hot-rolled surface. You can obtain both hot-rolled plates and hot-rolled sheets with various surface qualities. The difference in whether you have a hot-rolled plate, or a hot-rolled sheet is defined by the thickness. Plates are typically defined as a flat-rolled product greater than 4.6 mm (0.187 in.) in thickness.

The hot-rolled surface is not as smooth as the cold-rolled surface and the thickness and surface flatness tolerances are more liberal, due to changes in the surface as the hot metal cools.

Hot-Rolled Carbon Steel Plate

There are a number of surface quality designations in use when ordering hot-rolled steel plate. Hot-rolled steel plates are minimum 4.6 mm thick (0.187 in.). Hot-rolled carbon steel plates have a maximum of 0.33% carbon. There are various quality descriptions to be aware of when ordering steel plate for art and architecture. The most common and simplest surface is the *commercial steel* designation. It used to be called *commercial quality* or CQ.

Commercial steel (CS) quality is the most economical of the carbon steel plates. Tolerances of alloying constituents, mechanical properties, and flatness are not tightly controlled. Surface imperfections, scale, and stretcher strain marks may also be visible on regular-quality hot-rolled carbon steel plates. This is the designation for mild steels.

There are other quality descriptions used when ordering plate that require tighter tolerances in the hot rolled steel plate. Structural steel, SS, drawing steel, DS, and a number of others associated with industrial practices and requirements. Chapter 5 goes into more detail and cross references this to the older quality designations.

The quality of the surface is secondary for most mild steel. Commercial steel plate can have more surface scale and inclusions. CS sheet can have coil breaks, stretcher strains, and other surface defects. Mechanical properties have a wider range. If specific mechanical properties are desired, then more information should be needed and working with the supplier of the steel plate is critical.

For the commercial steels, drawing steels, and structural steels there is not any particular surface consideration. The surface quality from the mill is not of primary concern. Many of these steels are specific to a particular operation and thus a tighter control on the mechanical properties and less concern on surface appearance since most steels are treated and coated with an organic coating of some form or galvanized. There can be imperfections created from the coiling and decoiling processes, some minor oxide, scratches but usually scale is removed. Figure 3.5 shows a typical steel plate in the right image and structural steel beams in the left image. In most cases the surfaces are free of scale and heavy oxide.

FIGURE 3.5   Structural steel and structural steel hot-rolled plate.
Source: L. William Zahner.

The hot-rolled carbon steel surface is the finish you see on structural steel shapes before finishing. This mill surface will oxidize and corrode if left unprotected in a moist environment, so most steel structures have the scale removed, then coated with a protective layer of paint. The veneer you see on exposed structural steel has been prepared by blasting with abrasives to remove the scale and oxides from the surface. Then coated with a red or gray zinc primer paint coating.

Hot-Rolled Carbon Steel Sheet

The carbon steel sheet comes in two forms, hot rolled and cold rolled. This is different than most other sheet metal material where hot rolled is usually limited to thicker plate material.

Depending on the width of the sheet, carbon steel sheet can be hot rolled to as thin as 1.2 mm (0.045 in.). There is some overlap with what is usually considered plate in the higher end of the range. Hot-rolled sheet is typically a maximum of 6 mm (0.23 in.) in thickness before it is considered hot-rolled plate. However, for some large width sheets, thicknesses up to 12.5 mm (0.5 in.) are being produced by some steel mills using continuous casting methods. See Table 3.1.

The descriptions used on hot-rolled carbon steel sheet are dependent on the end user requirements. For use in art and architecture, the most common quality designation is *commercial steel* or CS. Commercial steel does not have tight tolerances on alloying throughout the sheet nor does it have tight mechanical properties. It is still ductile and can be pierced, stamped, and formed relatively easily.

TABLE 3.1 Thickness of typical hot-rolled sheet.

Minimum thickness	Maximum thickness	Width
1.2 mm	6 mm	300–1200 mm
0.045 in.	0.23 in.	12–48 in.
1.2 mm	0.045 in.	1200 mm and greater
4.5 mm	0.187 in.	48 in. and greater

The commercial steel hot-rolled sheet may have been given a temper pass to improve the surface quality. This is done when the finish surface needs to be superior for cases where the steel is exposed or painted. The commercial steel may be simply annealed to arrive at a specific temper. These instances do not have a need for good surface quality and may have more streaking and surface imperfections. In this case, the surface quality is lower than the temper pass surface.

Keep in mind, the surface quality is not normally subjected to the scrutiny of the art and architectural world. Steel is used widely in industries with less stringent surface requirements. Carbon steel sheet is usually destined for coated or galvanized parts used in less aesthetic industries. Unlike the aesthetic requirements a designer places on the surface appearance of stainless steels, aluminum or brass, the surface appearance of carbon steels often is not the primary concern.

Commercial steel hot-rolled sheet is subject to three different surface quality levels. If you simply order commercial quality hot-rolled carbon sheet you are going to receive what the supplier has on his floor. This will be in one of these three categories:

- As rolled
- Pickled and dry
- Pickled and oiled

The differences are significant and should be included in any specification when ordering carbon steel sheet. The "as-rolled" designation can have scale; it is usually darker in color and streaks of different color tones run the length of the sheet. Often, the edges are different in color tone. It can have oxide or rust from storage and transportation. The rust is superficial and limited, but in the event it is excessive, the sheet should be rejected. Significant rust is an indication of inadequate storage.

The pickled and dry sheet material is free of scale and surface oxide. It can have some streaking color tones running along the length. Because it is not oiled, some storage oxide, in the form of a light rust, can be apparent on the sheet.

The pickled and oiled has no scale and is oiled at the mill to keep it from oxidizing during storage and shipping. The oil is a mill oil and needs to be removed in a degreasing process. Mill oils are thin synthetic hydrocarbon oils that you can feel and will wipe off. They are excellent for keeping the steel from oxidizing but the interfere with subsequent processes and needs to be removed. Commercial degreasers will remove these oils, but this adds processing costs.

TABLE 3.2   Quality designations for hot-rolled sheet.

Quality designations
• Commercial steel
• Forming steel
• Drawing steel
• Deep drawing steel
• Extra deep drawing steel
• Structural steel
• HSLA: high-strength, low-alloy
• Solution hardened steel
• High-temperature steel

Other than commercial quality sheet, there is drawing quality and drawing quality specially killed. These are specified when deep forming and stamping operations are to be performed on the steels. These quality designations have more stringent uniformity requirements than the commercial steel.

As with the CS, each of these finishes is also subject to the additional qualifiers of "as rolled," "pickled and dried," and "pickled and oiled." See Table 3.2.

Of those quality designations, the most common in the art and architectural world are the commercial steel, forming steel, structural steel, and HSLA steels. The others are designed for specific industries requiring specific characteristics.

Cold-Rolled Carbon Steel Sheet

Cold-rolled carbon steel sheet has better surface quality characteristics when it comes to smoothness, lack of inclusions, scale marks, and geometry. Cold-rolled carbon steel is used across industries for a myriad of products. Thus, consistency, predictability, and quality are paramount.

Cold-rolled steel sheet is classified into two main categories, Class I, where the surface is intended to be exposed, and Class 2, where the surface is intended to be concealed. Similar to the hot-rolled steels, the cold-rolled steels have quality designation specifications.

The cold-rolled steels are provided in a given quality standard, depending on the requirements and end use. Note this is a North American designation and is part of the idiom that has surrounded steel production for better than a century. See Table 3.3 for a list of the designations used for the cold-rolled steels.

Commercial Steel

The commercial steel designation, CS, is a common designation for low-carbon steel, cold-rolled sheet. The surface achieved under this qualification for cold-rolled steel is superior than the surface

TABLE 3.3 Quality designations for carbon steels.

Quality designation	
Commercial steel	CS
Drawing steel	DS
Forming steel	FS
Structural steel	SQ

for the regular quality hot-rolled steels. The surface on commercial steel cold rolled is a smooth, matte surface free of scale and oxide. The steels provided with this designation are not as exacting in chemical makeup or mechanical properties as other quality designations. They are ductile and formable. They can possess internal strain that may require leveling as they move through manufacturing.

The surface of cold-rolled commercial steel is suitable for coating with paints or varnishes. This surface allows the chemical processes of oxidizing and darkening. The surfaces of cold-rolled commercial steel are free of streaking and linear inclusions.

Drawing Steel and Forming Steel

Drawing steels possess more precise characteristics because their intended use requires specific behavior as the metal is formed in stamping and drawing operations. The designation is DS. Chemical composition is not as tightly controlled, but the mechanical characteristics are defined by the end user. Drawing quality is more ductile and predictable than commercial steel; however, internal strains can still exist within this quality designation and may require leveling or other methods to reduce internal strain in sheet or plate material.

Forming steels are similar to drawing steels, with good ductility and formability. These are often designated to be galvanized.

There are other drawing steels, the deep drawing steel, DDS, and the extra deep drawing steel, EDDS. These have lower levels of carbon and are very formable. The EDDS has very refined grains due to the additions of titanium or columbium. The carbon content is very low, less than 0.005%. These are the most formable of the steels.

Cold-Rolled Structural Steel

The structural quality steel, SS, is designated when specific mechanical properties are required. Along with the SS qualifier, specific hardness or yield and tensile strength requirements should be included. Cold-rolled structural parts are always specified to be made from this quality designation. This designation is used when making light-gauge support members from cold rolled carbon sheet.

Channels, z-girts, and hat sections used as support members should be made from SS designated steels.

Similar to the hot-rolled steel sheets, these cold-rolled steel sheets are subject to three different quality levels:

- As rolled
- Pickled and dry
- Pickled and oiled

The surface finish on the cold-rolled steel is a lighter color and more refined than the hot-rolled surface finish. Today, the finish available on the cold-rolled steel surface is highly refined. This is due to stringent manufacturing process from across industries. The need for a predictable sheet steel surface with a consistent tight microscopic structure has been met by the steel industry.

The automotive industry is the largest user of cold-rolled steel sheet. Much of this is destined for stamping and finish coating. A predictable, consistent steel sheet is demanded by the rapid production processes, such as stamping and welding used to make the car body forms. Inherent flaws in the metal need to be culled out before they arrive at the manufacturing plant, so precision casting and cold-rolling process have been established to meet these demands.

The art and architectural use of cold-rolled steel is in a position to benefit from these quality procedures. When acquiring cold-rolled sheet, these same exacting processes have been performed from the original mill source.

There are several finishes levels available. These are imparted to the surface based on the nature and quality of the cold rolls. For most applications, the surface finish is a matte, dull reflective tone. Often, this is the base surface for paint coatings, and the texture and gloss are sufficient to provide a base for paints. Clear coatings can also be provided to the cold-rolled finish texture. Wax or clear acrylics are common coatings used to show the natural matte color of cold-rolled steel. Often, a slight grain is apparent on the cold-rolled sheet running the long direction. Today, much cold-rolled steel is produced with a tight and minimal grain appearance, nearly indistinguishable with the eye. This is due to these stringent controls at the mill where annealing and cold-rolling processes are in place to reduce the grain size and arrive at a more isotropic metal surface. Figure 3.6 shows an interior wall surface with natural steel, unweathered, and without a conversion coating. The surface is made from thin, hot-rolled carbon steel. Darkening along each edge is common with hot-rolled, continuously cast steel sheet. The alloy is most likely G10080 carbon steel.

Cold-rolled carbon steels can also be provided with what is called a "commercial bright" finish. This is a smooth, somewhat brighter finish than the matte. There is less grain apparent. This finish is used as a base for plating or fine painting applications. The surface is produced on higher-quality cold rolls.

Another finish rarely provided on low carbon cold-rolled steels is the "luster" finish. This finish is induced into the steel surface by rolling in fine polished rolls. The surface requires coating or plating as it will eventually roughen from oxide growth. Fabrication processes will reduce the luster and reflectivity.

FIGURE 3.6 Image of the natural steel look.
Source: Courtesy of Dan Gierer.

Mill Oil

On both hot- and cold-rolled steel, a light mill oil is often used to protect the steel surface during transfer and storage. This oil aids in resisting flash corrosion of the surface from exposure to changes in temperature as the metal is transferred from the mill source to storage and on to the end manufacturer. The oil is not meant as a lubricating oil. It has various corrosion inhibiting additives and is applied as the metal leaves the last set of rolls. These mill oils are specially formulated petroleum byproducts that fall into the family of mineral oils. *Mineral oils* are a broad term used to describe a number of different oil-like substances but for the steels, these are usually alkanes, a byproduct of gasoline production. The corrosion resistance is obtained from the additions of sulfonates into the oils. Sulfonates, such as sodium sulfonate, are used along with a surfactant.

This oil needs to be removed prior to finishing with a clear protective coating or prior to darkening or bluing. It leaves a slick, greasy feel to the touch and has a slight odor. This thin, clear oil has hydrophobic properties so it will not easily rinse off the surface unless a degreasing agent or detergent is used.

THE NATURE OF THE STEEL SURFACE FINISH

There are similarities between the steel surfaces and those of the copper alloys from the perspective of the artist and designer. For some designs, allowing the metal surface to age and develop a surface of strongly adhering corrosion products is similar to the natural green patina that forms on copper. Weathering steels, as an example, will develop a thick dark reddish brown coating over the surface of the steel, slowing the rate of change. The green color we see on copper and this orange to reddish brown on steel surfaces are composed of oxides developed by chemical and thermodynamic processes that provide significant corrosion resistance to the base metal.

Not all steels will perform like the weathering steels. In fact, the thought of what we call rust is generally an unfavorable outcome of exposed steel articles. Rust is considered an unwelcome and irreversible condition to be avoided by keeping moisture and oxygen away with the use of coatings. Rust is the enigmatic characteristic of steel material that for all other considerations is analogous for durability and strength.

Finish Possibilities Available for Steels

When we think of finishes on metals the satin finishes, mirror polish finishes, glass bead textural finishes, and others that are common on copper alloys, stainless steel and aluminum come to mind. For the steels, this is not the case. Steels can receive these finishes, even mirror polished, but the finish does not last. Firms that finish copper and stainless steels avoid finishing steels because it contaminates their equipment. Even glass bead blasting chambers have to be completely overhauled before stainless steel can be processed if steel has been blasted. Dedicated equipment is set to process abrasion of steels. If satin or nondirectional finishes are applied, it is usually to accommodate adhesion of enamels and lacquer finishes.

Beauty and durability can be accomplished with the special steels, the weathering steels, by forming a deep, rich oxide on the surface. These are the HSLA steels. But carbon steel can also be darkened, sometimes called *bluing,* by chemically treating the surface to create thin dark oxides of sulfides and selenides. Phosphate treatments can be applied as well to develop a thin layer of protection that affords a dark or nickel-like appearance. Note, however, that darkened surfaces and phosphate surface treatments require an additional protective measure of clear organic coating to withstand corrosion, while the weathering steel does not.

For steel surfaces, there is also a third option with different properties and appearances. Here the steel surface is bonded to another metal, usually in the form of a metallurgical bond between metals, as in hot-dipped galvanizing, where a layer of zinc is tightly bonded to the steel.

For the wrought forms of steel, plate, sheet, tube, and bar, various options are shown in Figure 3.7.

In considering each of these potential design paths for steel, the alloy and the microstructure of the surface are important. Carbon and other alloying elements in the steel that are dispersed across the surface will play into the final surface finish and the nature of the surface. Scale, the thick glassy oxide that can form on hot steel surfaces as it cools, will influence the appearance of the hot-rolled steel products. Even when removed by blasting or acid treatments, the steel surface can have indentations where the scale once resided.

Different appearances are obtained by the hot processes of hot rolling or hot drawing of the steel as opposed to the tighter microstructure developed from cold passes through cold rolls.

The first category to be explored, are the HSLA steel surfaces that form a natural, protective oxide. This oxide is developed over the exposed surface either naturally or by means of additive

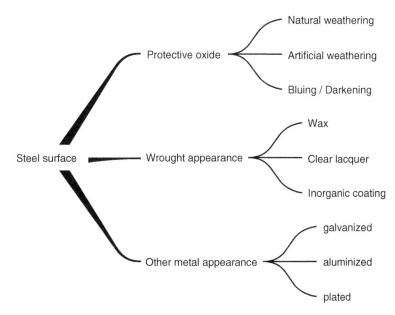

FIGURE 3.7   Surface choices for the wrought forms of steel.

chemical action to preweather the surface. These are called the weathering steels. Weathering steels are those sets of steels that form a tight, thick outer layer of an adherent crystalline oxide and hydroxide. These alloys of steel contain small quantities of copper, nickel, and phosphorus. An example of this finish is shown in Figure 3.8. Designed by Gensler Architects, this preweathered custom perforated steel staircase has the finish induced on all exposed surfaces.

FIGURE 3.8   Weathering steel surface used on a staircase. University of Kansas School of Business, Capital Federal Hall.
Source: Designed by Genlser Architects.

Protective Oxide – The Weathering Steels

The HSLA weathering steel alloys, as described in Chapter 2, are designed to develop a special thick iron oxide capable of slowing down the effects of atmospheric corrosion much like the green patina that develops on copper surfaces exposed to the atmosphere. This oxide grows on exposure to moisture and air.

The weathering steels were first developed by US Steel Corporation back in the 1930s. They were looking for a steel alloy for their ore carrying hopper carts they would use to transfer the iron ore and coal to the furnace. They needed a stronger, more durable steel alloy that had exceptional hardness and did not need to be painted. It was only later they realized this alloy had interesting corrosion resistant traits that could be marketed in other industries, so it trademarked the name COR-TEN®, often Corten, for this corrosion-resistant steel alloy.

The steel was not introduced as an architectural metal, however, until the 1950s. In 1964, the John Deere Headquarters, in Moline, Illinois, opened and was one of the first major architectural structures to be clad in weathering steel. Eero Saarinen designed the John Deere complex but, unfortunately, he died before seeing it completed.[1]

William Hewitt, then president of the John Deere company, said,

> The seven buildings should be thoroughly modern in concept but should not give the effect of being especially sophisticated or glossy. Instead, they should be more "down to earth" and rugged.

Saarinen chose this special weathering steel, Corten, to age and provide the earth tones and the "down to earth" feel his client desired. He knew the metal would age gracefully and create a deep, natural tone as the thickened oxide formed.

This intriguing metal, weathering steel, is sought for the timeless beauty of the oxide that develops on its surface. The deep rich surface oxide exhibits a natural brownish-red tone that is both durable and stable. Many artists have incorporated weathering steel into their work, the most well-known being Richard Serra. Serra has used weathering steel plates on numerous massive sculptures found around the world. Refer to Figure 3.9.

The great minimalist artist Donald Judd also used weathering steel in his works (see Figure 3.10). Judd often would preweather large plates and then allow them to continue to oxidize outdoors in Marfa, Texas. He also created smaller, weathering steel wall units where he worked the surface oxide to create a fine velvet-like appearance.

Since it was "discovered" as an architectural metal by Saarinen, weathering steel has been used on numerous outdoor sculpture and building facades around the planet. It has proven durability and resilience in many environmental exposures, while at the same time giving us a material that has a natural feel.

---

[1] Unfortunately, Eero Saarinen died before his other incredible structure, the Jefferson Memorial Arch in St. Louis, was completed.

FIGURE 3.9   Massive 50 mm thickness weathering steel plate sculpture. Private collection.
Source: Artist Richard Serra.

FIGURE 3.10   Weathering steel floor boxes. Sculpture by Donald Judd.
Source: Courtesy of the Zwirner Gallery.

The weathering steels are a special family of steels. They fall into the family of steels known as HSLA steels. Their alloying makeup is significantly different than the carbon steels. As with all uncoated steels,[2] the oxide develops when the surface is exposed to moisture and oxygen.

For the weathering steels, however, the difference lies in the nature of the oxide that develops over time. The oxide that gives weathering steel its protective ability is ferric oxyhydroxide, FeO(OH) an oxygenated version of ferric oxide – essentially, a mixture of oxides and hydroxide. This attractive, orange to rich purple brown oxide acts as a protective barrier to the base metal. When correctly formed, this tough layer of oxide, develops into a surface that resists atmospheric degradation in outdoor environments. Once the oxide develops correctly, further changes are very slow in most environmental exposures.

Weathering steel contains copper, phosphorus, chromium, and nickel as alloying elements. The small addition of copper along with the chromium, causes a change in the oxide formation at the surface. The initial oxide develops rapidly at first, a matter of days or weeks, depending on the moisture and temperature. Both are critical for the development of this initial oxide layer. Without moisture, as in a desert climate, the weathering steel may take months or years to form, if at all.

The development of the oxide is a chemical reaction on the surface. The chemical reaction occurs as the atoms of iron and oxygen collide. The more frequent these collisions the faster the reaction. The warmer the surface of the steel will increase the number of these reactions. This thermodynamic reaction is called the *activation energy* and the warmer the surface the greater the energy. The development of the iron oxide requires a surface temperature at least 10 °C (50 °F) along with oxygen laden moisture. The iron atoms go into the solution and collide with the oxygen. A surface that is too hot, however, will evaporate the moisture before it has time to create the ferrous ions needed to combine with the oxygen.

If the temperature range is right and moisture is present, the oxide will quickly form. This rich oxygenated layer develops when the surface undergoes a series of wet and dry cycles. Initially, a light layer of ferric oxide forms as depicted in Figure 3.11. This oxide is loosely established on the surface and is porous and brittle. It easily comes off onto one's hands or clothing. In a short time, if the environment is correct, it covers the entire surface with a soft, friable layer of oxide. In a way it is similar to what happens when brushing velvet. Brushing this friable oxide layer will leave a mark on the surface. This is the result of collapsing the microscopic peaks of oxide that initially forms. This initial layer of ferric oxide is responsible for the rust stain that forms below weathering steel surfaces that are allowed to form the oxide naturally. It is unavoidable during the weathering process. As the oxide develops it initially is loosely attached and will sluff off as moisture drips down the surface and grabs the particles only to redeposit them on adjacent materials.

---

[2]Stainless steels are the exception and are discussed thoroughly in the first book of this series.

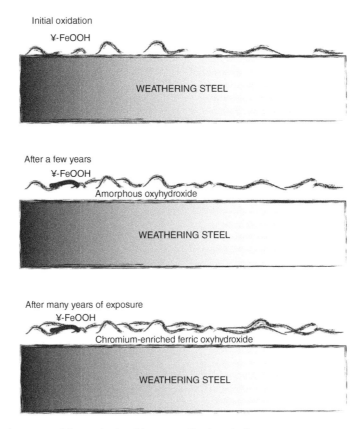

FIGURE 3.11   Development of the oxyhydroxide on weathering steel.

The initial oxide develops in "islands" across the surface of the metal. The color of these microscopic islands is orange-red and is made of thin crystals similar to the iron oxide mineral Wüstite. See Figure 3.12.

This oxide develops rapidly on the surface and goes through stages. Figure 3.13 shows a series of weathering steel roofs. These surfaces have been installed at different times. The orange red surface is approximately three weeks older than the roof surfaces in the foreground.

It is also worth noting, the roofs in Figure 3.13 that have not developed the oxidation appear as if they were made from standard low-carbon steel sheet. In actuality, the surface color in the unrusted state is similar for weathering steel alloys and other carbon steels. Even as the initial oxide develops, the two forms of steel look similar. It is only later, when the oxide thickens and darkens, that the real nature of the weathering steel surface become apparent. Regular carbon steel will flake away as the oxide grows, expands, and then sluffs off the surface. While with the weathering steel alloys, the oxide thickens and becomes dense.

On occasion, a design may wish the appearance of the weathering steel to stop at some intermediate point where the color is more deep orange than purple brown. The color can be striking.

FIGURE 3.12   Microscopic image of initial formation of oxide.
Source: L. William Zahner.

FIGURE 3.13   Roof surfaces initially installed.
Source: L. William Zahner.

Rich bright colors of orange and red first develop over the surface. Figure 3.14 shows the surface of the Trinity River, Texas, project designed by Antoine Predock. The surface color exhibits various stains or color differences, most likely initiated when the weathering steel was set out in storage on site. The growth of the initial ferric oxide develops from different levels of exposure. The lower right image is an art piece where this reddish orange tone is the desired surface. The art piece is displayed where humidity is limited and touching the surface is discouraged. The artwork will maintain the

FIGURE 3.14 Reddish-orange color tone develop during the early stages of exposure.
Source: L. William Zahner.

color tones since it will have limited exposure to moisture, while the exterior surface of the Trinity River project, exposed to the atmosphere, will thicken and darken.

As the surface is further exposed to moisture and oxygen, it develops into what is known as hydrous ferric oxide. The color is orange to reddish brown, and the surface is dusty because the layer of oxide is still in small, indistinct clusters as these "crystal islands" grow larger. In between these islands, ferric oxide mixed with hydroxide is appearing. Figure 3.15 is at this stage. It shows the microscopic nature of this surface. The oxide is thicker and more evenly dispersed over the surface. Darker layers composed of the oxyhydroxide are apparent with islands of the ferrous oxide as well. The surface is very crystalline.

The surface initially develops several iron oxide compounds. Table 3.4 lists a few of the common oxides of iron and the corresponding mineral form. Layers and islands of mixed oxides of iron oxide, ferric oxide, and ferrous oxide develop across the surface.

The copper in the alloy stabilizes the oxide and makes it more adherent to the surface. The initial oxide layer that grows is fragile and porous, not unlike what forms on regular carbon steel. As the surface continues through further wetting and drying cycles, the crystals form into dense, continuous clusters that become stable and intertwined to the base metal.

It is important to note that as the oxides and hydroxide develop on weathering steel, layering one on top of the other, until the more stable oxyhydroxide is established, the surface is fragile and will come off. This is also why the color and the surface texture is variable across the surface in the initial stages of weathering. Figure 3.16 shows a vertical wall weathering naturally. The image on the left is after a few weeks. The image on the right is after nine months.

FIGURE 3.15   Microscopic image of a weathering steel surface after a few months. The scale is 0.1 mm.
Source: L. William Zahner.

TABLE 3.4   Common forms of iron oxide compounds and their respective minerals.

Compound	Chemical name	Mineral	Associated color
FeO	Iron oxide	Wüstite	Red
Fe(OH)$_2$	Ferrous hydroxide		White to dark green
Fe$_3$O$_4$	Ferrous oxide	Magnetite	Black
Fe$_2$O$_3$	Ferric oxide	Hematite	Gray
Fe(OH)$_3$	Ferric hydroxide	Bernalite	
ΥFeO(OH)	Ferric oxyhydroxide	Lepidocrocite	Orange
σFeO(OH)	Ferric oxyhydroxide	Goethite	Dark purple-brown
σ(Fe1-x, Cr)OOH	Chromium substituted amorphous layer of Ferric oxyhydroxide	Chromium enrichened Goethite	Dark purple-brown color

Another aspect is the fact that the oxyhydroxide layer is insoluble. This insoluble layer develops away from the surface of the steel. As time and exposure continue, this surface thickens and bonds with the underlying steel. It develops on the more unstable ferrous oxide layer. If immersed in water, this important insoluble layer will not develop, Because of this, making fountains out of weathering

FIGURE 3.16   Natural weathering of a vertical wall.
Source: L. William Zahner.

steel is not wise. The surface will act as a typical steel surface and corrode away without developing the integral insoluble layer of oxide. It is this important oxide that gives weathering steels its long-term corrosion resistance.

Weathering steel oxidizes at different rates due to its orientation in the environment during exposure. Wet and dry cycles, exposures to rain and sun, will alter the color and the corrosion rate. The darkness of the weathering steel causes it to heat up when exposed to the sun. The absorbed heat dries the surface but helps to speed up the oxidation process. Generally, surfaces that are allowed to get moist but still dry out end up darker than those that face the direct sun and dry more rapidly.

Weathering steels' natural beauty is derived from the variations on the surface. The contrasting dark and light zones that appear sometimes within the same sheet or plate add to a mystique that few other metals offer. As with other naturally occurring surface colors, stone, wood, even the colors of the forest in fall, weathering steel gets its beauty from a natural oxidation process that occurs on the surface of the metal. Figure 3.17 shows stains and patterns of oxidation of a large, 7 m tall sculpture by Ewerdt Hilgemann. The right "dancer" is made from stainless steel.

As the surface exposure to the environment continues, crystals of iron oxide develop in closely packed zones on some areas of the surface and loosely packed zones on other areas. It is this variation that gives a richness to the way light is absorbed and reflected off the metal. If you want even, consistency in color and appearance then weathering steel may be the wrong material. It is not paint. It is a mineral substance that's very nature is one of diversity. It is a mélange of color derived from

FIGURE 3.17   "The Dancers" by the artist Ewerdt Hilgemann.
Source: artist Ewerdt Hilgemann.

the oxidation peculiarities of iron and other compounds that come from the surround environment. On the surface, various forms of iron oxide are apparent and the crystalline nature diffuse light and color.

Iron oxide develops in several phases. Refer to Figure 3.18 for microscopic images of various oxides growing on the surface of weathering steel. This means the oxide can exist on the surface of weathering steel in several mineral forms at the same time. The closest mineral form that develops on the surface is known as hematite and exhibits a rhombohedral crystal structure. There is also a cubic form that develop, and this closely resembles the mineral maghemite.

The process of moving through the stages of natural weathering takes time. The surface that develops is layered. The inner layer is an amorphous combination oxides and hydroxides often mixed with crystalline oxides similar to the minerals goethite or wüstite. On top of this layer is a partially ordered crystalline layer of ferric oxyhydroxide, $Fe_2(OH)_3$, which will restrict oxygen from reaching the base metal. When moisture is present, the copper and phosphorous in the alloy, dissolve along the cracks that form in the amorphous inner layer, creating small chemical reaction sites that generate more goethite-like oxide.

The outer layer that forms is a crystalline layer of oxides and hydroxides. This crystalline layer further impedes the movement of ferric ions outward and oxygen from moving inward to the base material. The lower left and right images in Figure 3.18 show this tight oxide development.

When allowing the surface to weather naturally, you should expect it to take three to five years, depending on the environment, for the oxide to thicken to the point where it is stable and sluffing off (*bleeding* as it is often called) slows. When the surface thickens, the color darkens to a reddish brown tone. During this period, expect staining from the runoff of loose iron oxide particles. Refer

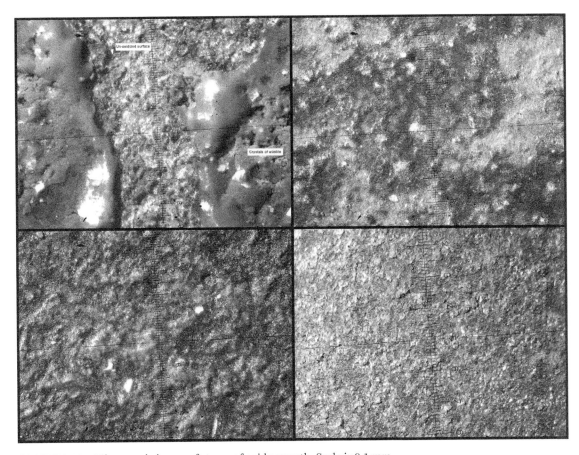

FIGURE 3.18   Microscopic image of stages of oxide growth. Scale is 0.1 mm.
Source: L. William Zahner.

to Figure 3.19. The white gutter liner shows the iron oxide runoff that happens early in the exposure and development of the final hydroxide.

In places where there are valleys or recesses, the oxide particles will collect and redeposit. These transition details will concentrate the particles and can make streaks on the metal and on the surrounding surfaces (Figure 3.20). These are temporary and will go away after the surface has been exposed for a period of time.

After five years, the surface thickens and darkens further. Staining diminishes as the oxide layer develops into the ferric oxyhydroxide. The color eventually reaches a deep brown to purple-brown color tone. Staining no longer occurs and the surface is very hard. Figure 3.21 is a 1.5 mm thick, roll-formed panel made from weathering steel. The surface has been exposed for nearly 40 years. Staining of the concrete sidewalk occurred for the first few years but has stopped completely since.

FIGURE 3.19   Runoff of the loose iron oxide particles. Private residence, designed by Marlon Blackwell.
Source: L. William Zahner.

FIGURE 3.20   Streaks from oxide growth variations on the surface of these corrugated panels.
Source: L. William Zahner.

FIGURE 3.21   Image of Zahner Bldg.
Source: L. William Zahner.

For many designs the idea of looking at a large expanse of corroding metal for weeks on end can be a daunting challenge for the average person to accept. Telling one to wait and allow nature to do its job may be too much for some people to tolerate. Preweathering the metal is often considered as an acceptable means of advancing the surface finish to an appropriate level.

Thin gauges, less than 1.5 mm, are not recommended because of the nature of the oxide development on all sides of the metal. The oxide that develops requires some of the "real estate" of the metal thickness. At first, a weak, loose layer of ferric oxide forms. Some of this comes off in the early stages of exposure. This soluble layer is what stains adjacent materials. This loose oxide also develops on the inner surface of the sheet and plate. Thus, for thin metal sheet, oxide development occurring on both the inner surface and the outer surface reduces the effective thickness of the metal and can perforate the metal. For thin weathering steel sheet, it is suggested that the back side of the metal sheet be coated with a zinc-rich primer or other barrier coating to prevent the dual surface oxidation that can occur. Even when preweathering the surface, on thin sheet, the reverse side should be coated in order to avoid premature deterioration of the metal from internal oxide development.

Preweathering – An Art and Science

Preweathering or artificial weathering of the surface of these high strength, low alloy steels involves exposing the metals surface to an oxidizing agent to start the process, then follow up with a series of wetting and drying cycles. True preweathering brings the surface of the steel to a dark, rich oxyhydroxide, with stable qualities similar to those that develop naturally over time. Once installed, staining should be minimal when preweathering is performed correctly. Most of the loose corrosion

FIGURE 3.22   Private residence in Telluride, Colorado. Designed by ARO Architects.
Source: Designed by ARO Architects.

product should be removed during the preweathering process. Figure 3.22 is of a preweathered surface where the staining of concrete elements below is minimized.

There have been many attempts to preweather the weathering steels for art and architecture. The artist Donald Judd supposedly washed his large plate weathering steel art pieces with muriatic acid, a dilute form of hydrochloric acid. This would cause immediate rusting of the surface. Rinsing the acid from the surface allowed oxygen and moisture to push the surface to a ferrous oxide, further rinsing and drying cycles would develop the deeper oxyhydroxide coating.

This method is strongly discouraged because of the fragility of the finish that forms. The strong acid pushes the development of the oxide rapidly and does not allow it to bind well with the base metal. Making a fragile oxide that easily rubs off. Not to mention, the dangers with working with strong acid and the requirement to have a method of neutralizing the acid as well as proper disposal. Any surface of any size would make this a particularly hazardous and environmentally destructive technique.

There are better techniques capable of arriving at a sufficiently adherent oxyhydroxide layer similar to that formed naturally over decades of exposure. These techniques are used on large architectural surfaces by firms that specialize in the procedures and have in place appropriate environmental and safety practices.

To accelerate the natural weathering process, you first must begin with a clean steel surface. The surface should be thoroughly cleaned with detergent to remove all mill oils, fingerprints, and process oils that may be on the surface. Rinse the steel in clean water, then follow with a light abrasion of the surface with an abrasive blast to remove scale and residual oxide. Follow quickly with a thorough wetting of the surface.

"Commercial blast cleaning" (SSPC-SP6)[3] of plate material will usually remove most mill scale and allow the surface to weather evenly. Blasting all the way to what is referred to as *white metal* (SSPC-SP5) can be performed for a very even appearance of the subsequent weathering of the surface. If performed manually, this can create mottling of the surface due to different levels and directions of blasting. These will translate through the oxide development, creating a lightly mottled surface as the oxide grows.

Initially, when you wet the surface there will be darkened runs and streaks that initiate from areas where the moisture is trapped or horizontal regions and where draining and drying is slower. It is recommended to gently rub these areas while wet with an abrasive pad or bristle brush. This will even these out by moving this ferric oxide across the surface.

You will need to allow the surface to dry and repeat this several times. It is the repetition of wetting and drying that weathering steel requires for the development of the sound ferric oxyhydroxide surface. The desire is to increase the amount of Fe^{3+} or Iron(III). This is the more stable form of oxide. You can augment the development of the oxyhydroxide layer by rubbing the surface with the abrasive pad or bristle brush while wetting. The layer of loose rust will come off during this process, and as it dries the surface will be more adherent and the color will begin to even out across the surface. However, this can take days if not weeks (Figure 3.23).

Acceleration and processing systems have been developed to create the surface in a controlled environment, an environment with humidity and temperature controls. The metal goes through cycles to generate a consistent, predictable surface and removes the loose particles that can stain adjacent surfaces. Figure 3.24 shows the semiautomated system of the Zahner Texas facility processing the oxidation to a designed level.

There are proprietary processes that involve accelerants to speed up the initial oxidation development. These are actually oxidation/reduction agents that create a condition where the surface is active and the oxidation of the iron on the surface is accelerated. It is this initial preparation of the surface that sets the weathering steel on a path to a deep rich surface finish. Many of these involve the use of chloride-based compounds. Some use pickling acids such as muriatic acid (dilute hydrochloric acid) sprayed or wiped on the surface, as the artist Donald Judd used. Handling acids and proper disposal of acids is not something to be done lightly. Subsequent wetting will generate a deeper color and arrive at the ferric oxyhydroxide surface desired. The acid will create a clean surface and actually removes iron oxide from the surface. The surface becomes very active and subsequent wetting rapidly develops the initial ferric oxide. The oxide that often develops from muriatic acid is loose and streaky.

---

[3]SSPC-SP6 is a specification for "commercial blast cleaning" process on steel. Appendix D describes this in more detail.

Initial prepared weathering steel plates

After 6 weeks of exposure natural and
accelerated wet and dry cycles

After 3 months.

FIGURE 3.23   Initial sand blasted steel plate, top image. After initial treatment and exposure, middle image. After several months, lower image.
Source: L. William Zahner.

There are several methods that do not damage the environment and push the oxide to the ideal preweathered surface. Figure 3.25 shows a surface created on formed plates for the Science Center at Amherst College.

This surface was preweathered by first blasting the welded and formed sections to prepare them for the initial oxidation treatment. The surfaces were treated immediately afterwards with a mildly

FIGURE 3.24   Preweathering plates in a controlled environment.
Source: L. William Zahner.

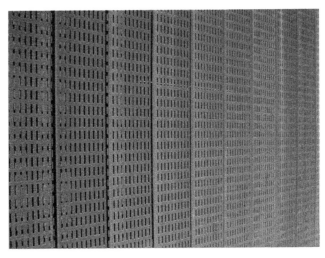

FIGURE 3.25   Pre-weathered steel plates perforated and formed for the Science Center at Amherst. Designed by Payette Architecture.
Source: Courtesy of Jeff Abramson AIA.

acidic treatment which started the oxidation process on the steel surface by creating numerous localized corrosion cells on the surface.

Once the surface reacts, the strong electrolyte is thoroughly rinsed from the surface. Preferably with deionized water. The surface is allowed to dry, then rewetted in continuous cycle until a thick, dark oxide is obtained.

The use of a sal ammoniac solution is sometimes used. Sal ammoniac is ammonium chloride. The ammonium chloride goes into solution and when applied to the steel surface becomes ferric chloride and releases hydrogen. This process is slower, but the results are adequate when followed by wet/dry cycling. Ammonium chloride is a corrosive salt often used as a fluxing agent on metals. It has to be removed in the subsequent rinsing or bright-yellow corrosion products will develop.

There are processes that utilize industrial-grade hydrogen peroxide as an oxidizing agent. Industrial-grade hydrogen peroxide is significantly more powerful than the 3–5% hydrogen peroxide available at your local drug store. Industrial grade is 15–30% and is a hazardous substance. This material decays into water and oxygen so disposal is not harmful to the environment, but it can be explosive if not stored properly. Industrial-strength hydrogen peroxide requires protective gear while it is being used. It will burn the skin and cause damage to your eyes and mucous membranes of the nose and lungs. Like the muriatic acid, it is not recommended as a substance to be used unless by someone versed in the use and control of dangerous chemicals.

The key for all the preweathering processes is first starting with a very clean surface, make it active by removing all oils, oxides, and other substances and then apply as many wet and dry cycles as possible. Roughening the surface by sanding or blasting increases the "tooth" and holds moisture to the surface. Eventually the surface becomes dark and stable. Subsequent runoff is reduced and the deep rich color is achieved. There are light and dark areas that add to the natural beauty but eventually you will achieve a deep purple-brown color across the surface. It takes time and repetition of wetting and drying cycles to achieve a good oxide that does not brush off or rinse off.

Several preweathering chemical compounds

Ammonium chloride	NH_4Cl	Mild acid in water
Hydrochloric acid	HCl	Powerful acid
Hydrogen peroxide	H_2O_2	Weak acid

The chlorine ion in solutions of ammonium chloride and hydrochloric acid is a strong oxidizer as it pulls electrons from the iron atom. Hydrogen peroxide is also a strong oxidizer. It provides oxygen to the iron. Combinations of hydrogen peroxide and the chloride ion work quickly on developing iron oxide on the surface by stripping the iron of electrons and providing oxygen to combine.

It is important, however, to understand that some of these act rapidly and form the oxide and hydroxide without binding well to the underlying steel. The surface may look the color and tone you want, but it will sluff off in powdery layers as it dries. Success comes when the process is allowed to slowly grow and there is diffusion into and out of the base metal.

For the sake of consistent and predictable results, there are propriety methods used to achieve the thick ferric oxyhydroxide surface. These methods are produced in a controlled environment where the humidity and temperature are regulated.

FIGURE 3.26   Preweathered steel plates used as walls for the "Jewel Box," Guggenheim Hermitage Museum, Las Vegas, Nevada.
Source: Designed by Rem Koolhaas.

As the steel is allowed to weather in the environmentally controlled chamber, the ferric hydroxide thickens, and the crystalline surface becomes less sporadic and more continuous. The layers are strongly bonded together. You cannot rub off the oxide. The color goes from an orange red to a deep purple red.

Preweathered steel plates, nearly 50 tons of 13 mm (0.5 in.) plate material, were used to create the walls of the "Jewel Box," a temporary exhibit on the Las Vegas strip (Figure 3.26). The plates were preweathered extensively to develop a sound oxide that would not come off when touched. The paintings were held to the walls using magnets.

Some of the ferrous oxide will still develop on preweathered surfaces when exposed to a moist environment. This is unavoidable because there is still porosity through the oxide layer. Over time, this porosity is reduced until a solid, impervious oxide has developed.

Weathering Steel – Hot-Rolled vs. Cold-Rolled Surface

The preweathered surface that is produced on a hot-rolled steel surface will be different than one produced on a cold-rolled steel surface. This is due in part to the microscopic irregularities on the surface and grain of the metal.

A hot-rolled surface will have more inclusions giving it a "tougher" appearance. This is due to the residual scale forming on the surface of the steel during the hot rolling process. The scale is usually removed during subsequent pickling processes. On thick plates, plates greater than 9 mm, the scale once removed can leave a scar or indentation. An example of scale is visible on Figure 3.27.

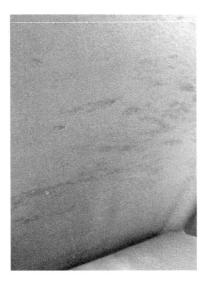

FIGURE 3.27   Scale on weathering steel plate.
Source: L. William Zahner.

The distinction between hot and cold rolled is centered on the process used to create various thicknesses of the sheet or plate. The manufacture of the sheet material is a subsequent operation to the hot-rolling operation. Essentially, the heavy hot-rolled plates are pickled to remove the scale and then passed through a series of reducing rolls that apply pressure and change the cross sectional geometry to thinner and wider sheet. These rolls will impart the temper into steel by cold working the metal ribbon as intense pressure is applied. What occurs is an elongation of the grains of the metal in the direction of the rolling and a smoothening of the surface as the smooth texture of the rolls is imparted to the surface of the steel ribbon. The texture of hot-rolled sheet and plate is less refined and rougher than the texture of cold-rolled sheet.

Cold-rolled weathering steel is available as thin as 0.8 mm (0.032 in. or 22 gage). However, this thickness should not be used on the exterior. It has a high potential of perforating from corrosion occurring on the back and front face. This has occurred on projects where moisture has access to the reverse side from condensation or other means. One solution is to paint the reverse side with a zinc-rich primer but still, this thickness, 0.8 mm (0.032), is too thin to be considered for exterior applications.

The maximum thickness for cold-rolled weathering steel is 4.76 mm (0.187 in.). After that, the metal is hot rolled. Hot-rolled weathering steel may receive a skim pass through sets of rolls similar to cold rolling. Passing through the rolls imparts a smooth surface onto the hot-rolled plates.

In the raw, unoxidized form, the weathering steel grades and the low-carbon steels are indistinguishable by simply examining the metal surfaces. Chemical analysis of the surface by means of an

X-ray fluorescence (XRF)[4] will determine the composition of the steel if you need to be certain that the right type and grade of steel is being used.

When considering weathering steel for the exterior of a surface, the choice between hot-rolled plates or cold-rolled panels or shingles is a matter of taste and design. Thinner, cold-rolled weathering steel allows for bends and returns that are sharper. Reveals can be expressed with a folded sheet thickness. Hot-rolled plates, on the other hand, provide a substantial, monolithic form that can withstand impact. Figure 3.28 shows both the thin cold-rolled sheet, upper two images, and the thicker hot-rolled plate, lower two images.

FIGURE 3.28 Thin cold-rolled preweathered steel on the top images and thick, hot-rolled, preweathered steel on the lower two images.
Source: L. William Zahner.

---

[4]XRF are X-ray fluorescence devices with the capability of reading the elements within an alloy and quantifying the percentage found on the surfaces.

The Controlled Surface

When it is desirable to have a very refined surface on the weathering steel, one that almost resembles the texture of velvet or worked leather, the outer veneer of the oxide can be burnished. This added process takes down the initial layers of oxide and creates a rust red slurry over the surface. As this slurry is allowed to dry, then lightly wetted and reburnished, the color begins to take hold. Loose particles are removed and a richness in color and texture sets in. These artistic surfaces perform best in the interior world where the environment is more controlled, and the surfaces are not continually wetted.

To create these forms of the surface, fine care must be extended by the artisan working on the metal. In addition, the surface created this way is more fragile. The iron oxide crystalline surface that forms is a *chromophore*. A chromophore is a group of atoms that are responsible for inducing color into substances. In the case of iron oxides, iron is the chromophore where in the Fe^{2+} state it can be colorless to green, while in the Fe^{3+} state, it can be yellow to orange red in color. What develops is the Fe^{3+} state, and the surface looks orange. Figure 3.29 shows a fine, velvety orange surface that has developed on a plate of weathering steel.

As the developing oxide takes shape during the preweathering process, the surfaces are hand worked while wetted not to the level of burnishing but to the point of breaking up this initial oxide development.

FIGURE 3.29   Fine, orange color and velvety texture on weathering steel plate that has been slowly worked to produce the finish.
Source: L. William Zahner.

The application of steam helps in this process by providing heat as well as moisture. As the oxide forms, use an abrasive pad or bristle broom to sweep the surface. As it dries, a very consistent, even appearance develops. It needs to be left untouched; otherwise, the soft ferric oxide crystals will smear around. If this happens, simply repeat the process. You can follow up with repeated steaming applications. These will thicken the oxide and improve durability. You can follow with a clear coat. The clear coat will darken the resulting finish considerably as it alters the way light is trapped in the oxide. The coating will help bond some of the loose fragments to the steel surface.

Interior Uses of Weathering Steel

The major constraint for using weathering steel as an interior surface material lies in the adherent nature of the rust, or lack thereof. The fear of it rubbing off onto one's clothes when brushed against the surface is a limit of its acceptability as an interior finishing material. The coarseness of the surface also makes weathering steel difficult to clean with conventional cotton rags as the threads hang up and leave small, difficult-to-remove fuzzy remnants on the surface.

There are two ways to overcome this issue. One is to coat the preweathered steel with a clear lacquer to seal over the surface. This reduces the roughness and will darken the surface considerably. Applying a hard wax over the lacquer surface will improve the cleanability of the surface. A good wax will hold up well over time and simplify the maintenance by arriving at an easier system to replace once the wax ages.

Another method is to work the surface similar to burnishing. The weathering steel will darken and grow the reddish brown tone as the loose particles are removed. Performing this several times will arrive at a steel with a dark, smooth texture that will not come off on one's clothing or when rubbed by hand. If moisture is allowed to set on the surface, however, it will form the orange color tones of newly developed oxide.

To arrive at a leather-like surface mist water and work lightly with an abrasive pad. Allow to dry. This procedure is repeated until the appearance has dark and light tones of oxide. Allow to dry and seal the surface. Reference Figure 3.30.

Constant working the surface as the oxide grows can achieve a surface with a great deal of consistency and beauty. Even large plates can be treated this way. Plates will have the coarser texture common with hot-rolled steels, and some darkened streaks may appear through the lighter red tones.

Galvanized Steel Surfaces

Galvanized steel is a ubiquitous material in common use for small utility items to large structural forms. Galvanizing protects steels with both a barrier coating and a sacrificial anodic coating. Produced by immersing steel in molten baths of zinc for a select period of time. The steel surface interacts with the zinc and some of the iron diffuses into the zinc at the interface of the two metals. Nearly pure zinc makes up the balance of the surface. Galvanizing produces a coating with one of

FIGURE 3.30   Preweathered steel. Leather-like surface created from working and burnishing the finish.
Source: L. William Zahner.

the lowest life cycle costs among all coatings. In the hot-dipping process of large steel forms, you can expect a service life of the coating of as much as 50 years.

Initially, the zinc is very shiny and reflective. Figure 3.31 shows a thick steel plate that has been newly galvanized along with a thinner plate showing the reflective surface and characteristic spangle.

On exposure to atmospheric moisture, the zinc coating forms zinc oxide, which darkens the surface. As further exposure occurs, carbon dioxide in the air forms a thin layer of zinc carbonate that darkens the surface further. These two oxides provide the long-term corrosion resistance characteristics of the metal.

In the galvanizing process for large steel structural shapes, the steel is first immersed in pickling bath to remove the surface oxide. This enables the molten zinc to create a metallurgic bond with the steel at the surface interface. When this occurs the zinc and steel intermix, and the zinc is firmly

FIGURE 3.31   Newly galvanized steel plate.
Source: L. William Zahner.

attached as if it is part of the steel. Typically, when a steel surface has been galvanized, it is not painted. Paint has a difficult time adhering to galvanized surfaces. When painting is necessary, the galvanized surface must be specially treated to enable the paint to adhere. Adhesion initially will not be the challenge it is over time as the zinc surface can go through a change and release the paint. Wiping the zinc surface with a mild acid, even vinegar, will assist in the adhesion of paint to the surface.

Protective Oxide – Bluing and Darkening

The process of darkening steel goes back centuries. Steel armor worn by knights in the fifteenth and sixteenth century often was darkened (Figure 3.32). Most likely, they used compounds of sulfur to produce a thin layer of insoluble ferric sulfide on the surface. The armor plates would be etched with intricate designs, then darkened with these sulfur compounds. When the steel surface was sufficiently dark, the armor would be polished to create highlights and contrast. Sometimes the etched regions would be infilled with gold or silver. This would create an even more dramatic appearance as the raised steel portion is polished and brightened, offsetting the silver, gold, or black infill.

These suits of armor were hot and cold worked to create the shapes and textures for the armor used to protect the aristocracy of the time. The suits were often commissioned, and elaborate decoration was hammered and etched into the surface. Perhaps the most amazing aspect is the level of detail this steel armor possessed, a clear indication of the craftsmanship and skill of the

FIGURE 3.32   Suit of armor. Etched, darkened, and plated steel breast plate.
Source: L. William Zahner.

time. The decorative treatments to the steel, followed much of the fashion of the day with slashings and pleats set in the metal to mimic the clothing of the aristocrat.[5]

Darkening of steel surfaces rose out of the need to protect the steel from corrosion as well as enhance the appearance. Suits of armor, muskets, swords, and other steel or iron features were expensive. It most likely began with oiling the surfaces or coating them with tallow. It was possibly determined corrosion could be hampered if the oxide was allowed to remain on the surface. These suits of armor were protected with beeswax, olive oil, and sheep's wool, which contained lanolin. Lanolin is an excellent rust inhibitor, and wrapping the armor in wool would offer some protection from oxidation.

Darkening or blackening of steel used to be achieved by applying oil to the surface and then heating the low-carbon steel in air or steam. The result was a low-gloss, darkened surface of variable thickness. It was labor intensive and more suited to small objects.

Bluing of Steel

Today, the process commonly known as bluing or gun bluing is a chemical process where the iron component in the steel reacts. There are two processes in common use:

- Hot alkali process
- Cold copper selenium process

---

[5]Krause, S. (2017) *Fashion in Steel*. Yale University Press.

In the hot alkali process, the clean steel is immersed in a heated bath containing sodium hydroxide and sodium nitrate or potassium nitrate. The caustic solution is heated to boiling and the steel article is placed into the solution for approximately 10 minutes. The result is a black layer over the clean steel surface. The black layer that develops is Fe_3O_4, which approaches the mineral magnetite. The coating is thin but is a passivating conversion coating. It has porosity, so allowing moisture to set on the surface will cause rust to form. Steels protected in this way are usually coated with a thin hydrophobic oil to repel moisture. These coatings are extensions of the steel oxide and are very adherent.

It is never easy or safe working with boiling solutions of sodium hydroxide, so another method was developed using a cold solution that creates a copper selenide layer intermixed with the iron oxide. These solutions are mostly proprietary, and working with selenium can also pose a hazard and must be disposed of appropriately.

This process involves thoroughly degreasing the surface and then dipping or wiping the copper selenium solution on to the surface. Usually a surfactant or etchant in the form of an acid are included to cause a reaction to occur on the surface. These cold solutions act very rapidly to darken the steel. The colors produced are dark blue-black, and the thickness of the coating is at most 10 μm. Similar to the alkali black oxide, darkened steel made this way should be coated with an oil, wax, or clear lacquer to prevent moisture from reaching the surface. The coating that develops is very adherent. Figure 3.33 is an example of blackened steel with a highly marbled surface.

The corrosion protection provided by these finishes is minimal. The hot process offers somewhat more in this regard, than the cold process. These coatings were created for the firearm and small tool marketplaces to provide low reflective surface that would give some protection. Producing them on larger sheet or plate can be challenging, but the results are dramatic.

FIGURE 3.33   Custom darkening of steel.
Source: Courtesy of Imagewall.

Another blackening or darkening solution involves phosphate salts. These salts use zinc or magnesium phosphate to develop a conversion coating on the steel. Similar to the hot alkali bath, the process of producing these coatings on the steel involves immersion in a hot bath, usually around 80 °C, (150 °F). This process is called *Parkerizing* and is used on mostly smaller parts. The iron phosphate surface is coarser, more crystalline than the other darkening processes. Abrasion resistance and corrosion resistance are improved. The phosphate coatings are dark gray rather than blue-black. They can be darkened by treating the iron phosphate coating with proprietary acid solutions or adding bismuth or antimony salts to create a smut that becomes absorbed into the pores of the phosphate. These coatings are also sealed after developing on a steel surface.

Blackening of Galvanized Steel

Zinc itself can be darkened by chemical treatments to the surface, therefore darkening newly galvanized surfaces is also possible since the outer layer of hot-dipped galvanized is of high purity. The process is somewhat proprietary in how the surface is initially prepared. You do not want to damage the protective galvanized coating and cause the surface to rust. Figure 3.34 shows the Diesel Store

FIGURE 3.34   Diesel Store in Chicago. Darkened galvanized steel.
Source: L. William Zahner.

FIGURE 3.35   Blackened galvanized steel front on Morimoto's restaurant.
Source: Designed by Tadao Ando.

with a tough, darkened galvanized steel surface. The white around the edges is due to zinc hydroxide forming.

Producing the darkening for use out of doors is very difficult because you do not want to do too much damage to the protective zinc surface and cause the underlying steel to corrode. The chemical action involves creating a dark zinc carbonate on the surface intermixed with other proprietary chemistry in such a way that you extend the lifetime of the galvanized layer. If you can make the zinc not want to react with the environment, you leave it to protect the steel if the steel is attacked. Figure 3.35 is a hot-dipped galvanized surface darkened for the Tadao Ando designed Morimoto's restaurant in New York City. This has been in place for over 10 years and has performed very well.

Color Tinting of Steel by Heat

Steels can be heat tinted to create interference oxides on the surface. This is not normally used in architecture, but there are artistic creations on small articles. The process is called *tempering* or *heat tempering*. The difficulty lies in heating the object uniformly then when the color is achieved, quench the steel to freeze it. The surface must then be dried and coated to prevent corrosion. The colors go from silver to gold to red on to blue as the temperature increases (Table 3.5). See Figure 3.36.

Chromate Treatment

Chromate treatments are conversion coatings that aid in passivating the steel surface. It is an excellent corrosion inhibitor and its primary use is as pretreatment for metals.

Applying chromate pretreatments to steels is not a direct process. It usually involves plating the steel first with zinc. Chromate treatment is a common coating to small steel objects that need some

TABLE 3.5   Temper colors and approximate temperatures.

°C	°F	Color
175	350	Light gold
200	400	Gold
227	440	Bronze
260	500	Red
282	540	Purple
310	590	Blue
343	650	Light blue
388	730	Pale blue

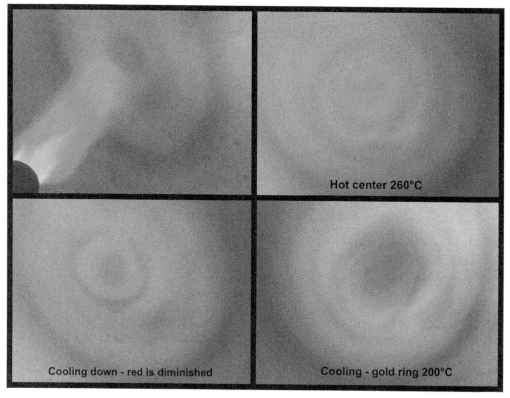

FIGURE 3.36   Image of heat tempering.
Source: L. William Zahner.

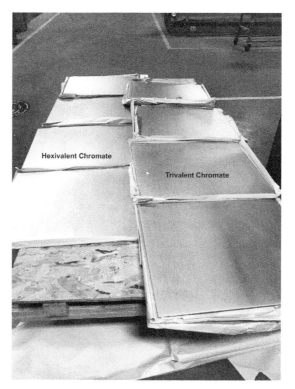

FIGURE 3.37   Chromate treatments on steel. The left side is the more iridescent hexavalent and must be coated if it is to be extensively touched or handled.
Source: L. William Zahner.

level of surface corrosion protection. It is often used as a standalone coating for steels because of its attractive golden/greenish–yellow, iridescent color.

There are two forms of chromate pretreatments, hexavalent and trivalent. The hexavalent form is highly toxic and highly regulated (Figure 3.37). The trivalent is less toxic but gives a slightly different color. It is highly advised the hexavalent be coated in a clear coating to seal the surface if these are to come in contact with humans. It has over the years been used extensively as a corrosion inhibitor on steels and aluminums. There have not been enough studies to determine if the coating, when hardened, will lead to any issues of health concern.

The coating systems used are generally proprietary, but the basic approach is a dipping process. After zinc plating, the steel is immersed for several seconds, then force-dried at low heat.

Painted Steel Surfaces

Most steel surfaces used in art and architecture are coated with organic or inorganic coatings with the exception of the weathering steels. Most paints were designed specifically with steel in mind.

The key to painted surfaces, beyond the color desired, is the preparation of the surface to receive the paint coating. The steel surface must be free of all oils and grease. Slag left on the surface, once coated, may flake off, exposing the raw unprotected steel.

Corrosion inhibitors are suggested for retarding the develop of the electrochemical interactions of the steel and iron surface with oxygen. Many of these do this by inhibiting the anodic polarity or cathodic polarity from developing. The inhibitors in use today are:

Phosphates

Benzoates

Nitrides

Aromatic organic materials

These can pose a hazard in handling and disposal. Use should be limited to those firms with knowledge of how to handle and dispose of the waste. They need to be compatible with the coating system used to provide both the protective outer coating as well as the decorative finish appearance. But before anything is applied to the steel surface, it must be clean and free of all oils and grease that will interfere with the adhesion of the paint or primer system.

SURFACING EFFECTS

Damascene and Pattern-Welded Steel

Damascene is a decorative effect on steels. Damascene relates to contrasting lines that appear to layer over the surface of the metal similar to the way the grain appears on a cut and sanded wood surface. The lines do not cross over one another but appear as frozen contour lines on a drawing.

The terms *damascene* and *pattern welded* refer, for the most part, to the same process, damascene being the eastern name for the process and pattern welding being the western name. They both involve the decoration of the surface by the microscopic alternation of steel types. Another descriptive name sometimes given to the metal is *watered steel*, as the contrasting colors resemble the marks in sand left by waves hitting a beach. By varying the steel types and repeatedly hammering, twisting, and shaping, this ancient technique produces some amazingly beautiful surfaces. Figure 3.38 are images of modern-day damascene steel plates that resemble those of the earlier work.

Ancient techniques were used to produce swords and knife surfaces that possess an organic, flowing texture called *hamon,* from the Japanese word for the distinctive sword blade texture or *damascene* for the beautifully intricate lines found on ceremonial swords of the Persian, Arab, and Indian cultures (Figure 3.39).

These beautiful surfaces are created by layering thin strips of iron, heating, and caburizing,[6] and hammering repeatedly to weld them together by pressure. This would be repeated as the pieces are again heated, folded, and hammered. Steel, unlike other metals, can be joined together by intense

---

[6]Carburizing is the technique of exposing hot steel or iron to a carbon-rich atmosphere. The steel on the surface absorbs some of the carbon and forms a hard, carbon-rich layer on the surface.

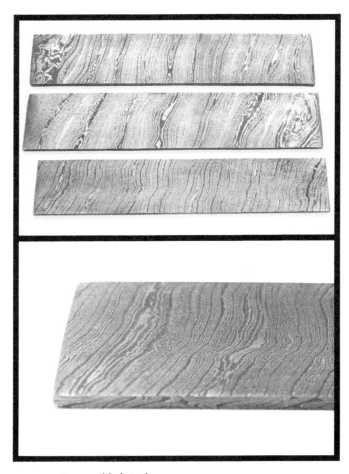

FIGURE 3.38   Damascene or pattern-welded steel.
Source: L. William Zahner.

FIGURE 3.39   Knife blade made from pattern-welded steel.
Source: L. William Zahner.

pressure from hammer blows. The oxide intermixes and the metal binds together. Copper oxides, on the other hand, inhibit the joining of two layers of metal by hammering.

The process uses the segregation of the microstructure of the steel to enable a contrasting pattern generated as the metal is folded and shaped. The ancient blade makers using this technique understood that different characteristics could be achieved by altering what happens to the steel. They may not have understood the metallurgic behavior, but they did know they could achieve different characteristics by altering how the steel is treated. Some steel could be hardened so a sharp edge could be achieved. They realized that other processes would lend a flexibility to the steel so it would not fracture from brittleness. What they were achieving was a mixture of high-carbon steel with low-carbon steel layers, each with different characteristics of hardness, color tone, ductility, and corrosion resistance. The corrosion behavior allowed some steel at the surface to etch differently than other steel. These ancient swordsmiths could combine steels to arrive at these characteristics in the same sword blade. The added effect of the beautiful contrasting lines added to the mystery and, most surely, the desire for this steel.

The ancient swordsmiths used several techniques to create the sword blades. Layering alternating strips of iron and steel and then heating and hammering them was a common practice. Other swords, perhaps when time was of the essence as when an advancing horde was approaching, were produced by laying down iron strips over a thin strip of steel, so the steel protruded to the edge. Then hammering them hot to weld them together.

A late Saxon period sword, discovered in England, was extensively examined by the Reading Museum in Berkshire, England. This long blade was found to be composed of layers of iron and steel, 12 in total. The cutting edge was steel containing differing amounts of carbon, most likely due to heating and reheating. The core was a made of several layers of iron.[7]

The Freer Gallery of Art has a collection of various patterned steel blades dating back centuries. Several of these were tested to determine the makeup and composition of the steel. XRF analysis, a nondestructive surface analysis used extensively on metals, indicated there are trace amounts of nickel on the surface. This would indicate there is a possibility of the iron used to manufacture these places, may have come from a meteorite collected on the surface.

Today pattern welding uses alternating layers of high- and low-carbon steels to instill different appearances and characteristics to the surface. Several different layers of iron and steel are welded together, hammered, twisted, folded, and rehammered, not unlike what was performed by ancient metal workers of the third century except today with more modern equipment. It is still a slow and limiting process, more suited to art and craftsmanship than large-scale architectural production.

The distinctive contrasting color bands can be brought out by etching the surface with mild acid solutions. Lemon juice, vinegar, solutions of copper sulfate (known as Vitriol), and mild sulfuric acid will etch the surface of the steel and bring out the contrasting color bands that give the steel the character of damascene steel.

The creation of these patterns is an art form and a practiced skill. The flowing worm-like patterning created by a skilled artisan creates a beautiful enigma on these steel surfaces. To produce this on large surfaces requires a completely different tact. Steel sheet and plate are far more consistent today than the small strips of steel made in the foundries and metalsmith shops of old.

---

[7]Scott, D. and Eggert, G. (2009) *Iron and Steel in Art*. London: Archtype Publications Ltd.

There are several means of producing decorative surfaces over steel plates. Applying a chemical resist and chemically etching the surface with various substances is one method. Steel can be selectively dissolved using various chemicals, all of which need special handling and disposal systems in place. The challenge that develops is protecting the steel surface after treatment. Any exposure to moisture can begin the oxidation process and seeds of rust will be established. Thus, a method of quickly drying and protecting the surface is needed.

The larger constraint is scaling the production to accommodate even a table-size surface. There are ways of achieving decorative surfaces, but they do not have the same depth and mystic of the damascene steels on small sword or knife surfaces.

Laying Down Weld

Laying down welds of different steel alloys and oxidizing the surface is another method in use today. See Figure 3.40. In this image, weld is laid down in lines that create a texture over the surface. The

FIGURE 3.40   Laying down weld to create a texture.
Source: Artist R + K Studios.

technique, still somewhat limited in scale, creates differing surface morphology of the weld metal itself and the diffusion of some metal from adjoining areas into the weld.

For the most part, these techniques are in the realm of artists, and scaling the processes up to architectural surfaces are not possible. For large surfaces, chemical variations on the steel surface can be a better means of creating the contrasting tones.

Carved Steel

Another steel surfacing technique, also in the realm of art more so than architecture is a process of "carving" steel with a torch. Torch carving heats the steel with an acetylene torch, and as the metal liquifies on the surface, it is blown away by the torch gases. This creates an amazing surface texture. The steel plate shapes as heat is applied and the metal is "carved" from the surface. Figure 3.41 shows a series of steel plates that have been carved and oxidized. The top left is the incredible surface

FIGURE 3.41   Carved steel artwork by R + K Studio.
Source: artwork by R + K Studio.

FIGURE 3.42   Ring of hammered steel.
Source: L. William Zahner.

generated by the artist and the hand torch. After carving, the surface is coated with a clear coating to protect from corrosion.

The process is very slow and difficult. It has the potential to be optimized by adding robotics to the cutting movement, but the result is not the same as what an artist adds to the work. Figure 3.42 shows another steel surface that has been heated and hammered to create a unique artistic texture.

Steel is an amazing metal in the hands of an artist, and the finish possibilities are endless. The color, feel, and texture of the metal has interested humankind for centuries, and it is likely this adventure will continue as new technique and artistic interest accelerate.

CHAPTER 4

Expectations

Vladimir   : *That passed the time.*
Estragon : *It would have passed in any case.*
Vladimir   : *Yes, but not so rapidly.*

Source: Waiting for Godot, by Samuel Beckett. © *1953, Grove/Atlantic, Inc.*

INTRODUCTION

Ferrous parts have a considerable affinity for oxygen, which is one reason why pure iron is not found on Earth's surface, only as compounds of iron and other substances, mainly oxygen. Iron makes up a considerable amount of the earth—nearly 5% of Earth's crust and a significant portion of Earth's mantle. Earth's core is solid iron surrounded by liquid iron and a layer of nickel. The earth is essentially a giant iron ball.

There are those occasions where iron fell from the sky in the form of a meteorite. Early mankind found these high-purity sources rich in iron, a gift from heaven, they thought. Greeks called iron *sideros,* which meant "from the stars."

Today we obtain iron from smelting various ores such as magnetite and hematite. See Table 4.1 for a list of the major ores from which iron can be acquired. Figure 4.2 shows some of the beauty these mineral forms can take. Producing iron from its ores demands a lot of energy to free the iron. This energy is essentially stored in the iron and finds release by combining with oxygen to form that ubiquitous substance we commonly refer to as rust. See Figure 4.1.

Rust is ferric oxide often hydrated and mixed with various iron hydroxide compounds. The closest mineral forms are Goethite and Limonite, $FeO(OH) \cdot H_2O$. The color of Limonite is orange to reddish brown, similar to the color of rust. As iron hydrates, ferrous hydroxide and ferric hydroxide

TABLE 4.1 Common minerals of iron.

Mineral name	Formula	Color
Magnetite	Fe_3O_4	Black with a brown tint
Hematite	Fe_2O_3	Gray to red
Goethite	$FeO(OH)$	Dark brown to black
Limonite	$FeO(OH) \cdot H_2O$	Yellow to orange
Pyrite	FeS_2	Brassy yellow

FIGURE 4.1 Rusty ship.
Source: By Shutterstock.

form on the surface, usually intermixed with the oxides. Ferric hydroxide can have colors from yellow to dark brown, even black while ferrous hydroxide has a greenish tone.

Rust is also a name for a color. Stage lighting in theater have a lighting scheme often used to produce the color tone of rust. Mix yellow and blue to arrive at green, add red, then more yellow, and you have rust. In 2018 it was a fashionable color for clothing. Then in 2019 it lost out to "burnt

FIGURE 4.2   Minerals of iron. Goethite, hematite, and pyrite.
Source: L. William Zahner.

orange." There is also "burnt sienna" in the Crayola™ crayon box. Rust must not be sexy enough to warrant a crayon.

When we think of steel and iron, it is hard not to think of the corrosion product at the same time. Rust is one of the most natural occurrences that happens on this planet. Entire industries have strategies that include fighting back the effects of rust. It is like gravity, always present. We need to find ways to deal with it.

THE DEVELOPMENT OF RUST

In reality, when we see rust forming on steel or iron surfaces, it is a combination of several forms of oxides, hydroxides and oxyhydroxide. Iron requires something to trigger the formation of iron oxide. Moisture is the common trigger for the development of the oxide, and it is not necessary to have ample amounts of moisture; simply moisture in the air will create the trigger. If the iron is kept dry, the oxide is slow to develop, but time and thermodynamic drive will eventually form a layer of oxide.

When moisture reaches a ferrous surface, some of the iron will dissolve into the water in the form of iron cations, positively charged atoms, Fe^{2+}. In Chapter 1, the iron atom is shown with two electrons in the outer shell. The pair of electrons flow to the surrounding areas, creating a tiny electrical current. When this occurs, an anodic region develops around these positive charged atoms while surrounding areas become cathodic.

The water molecule will separate into H^+ and OH^- partially on the surface. Additionally, there is oxygen dissolved in the water so the availability of the negatively charged hydroxide, OH^-, to combine with the positively charged iron cation now in the water solution makes for a rapid development of ferrous hydroxide, $Fe(OH)_2$. As more oxygen is introduced, either from the dissolved oxygen in the water or from oxygen in the surrounding air, ferric hydroxide forms, which converts partially to the mineral form, hematite, iron oxide, Fe_2O_3.

The chemical reactions that lead to rust formation are:

$Fe \rightarrow Fe^{2+} + 2e^-$	The iron goes into solution as a cation
$H_2O \rightarrow H^+ + OH^-$	Water disassociates into an ionic form
$2H_2O + O_2 + 4e^- \rightarrow 4OH^-$	More hydroxide develops at the cathode region using the free electrons
$Fe^{2+} + 2OH^- \rightarrow Fe(OH)_2$	Ferrous hydroxide forms initially
$Fe(OH)_2 \rightarrow Fe(OH)3$	Some of the ferrous hydroxide forms ferric hydroxide, and this, in turn, will form ferrous oxide

For the steel surface, keeping it dry is imperative or expect the development of oxides and hydroxides. The drive for oxygen is powerful, and even interior spaces can have sufficient humidity to eventually lead to the appearance of a red oxide on the surface. Of course, if the idea is to develop the oxides on the surface, as with the weathering steels, it still may be advisable at some point to restrict the surface exposure to moisture.

CHOICES AVAILABLE TO THE DESIGNER

The use of ferrous surfaces in art and architecture requires the designer to consider several options, as shown in Figure 4.3. These options are based on the appearance the design is striving to present. If the path is for the use of carbon steels with a natural as rolled or cast surface, the options are limited. Each will require some level of protection by sealing the surface so that oxygen cannot reach the iron.

Another option is to create a compound of iron oxide such as Fe_3O_4, similar to the mineral magnetite on the surface.

The process of darkening the steel surface can be done in several ways; hot treatments in phosphate salt baths, hot alkali baths or cold treatments to arrive at a thin layer of copper selenide on the surface. These treatments provide some oxidation protection by creating conversion coatings on the surface. These conversion coatings add oxidation protection to the steel; however, eventually there will be some diffusion of iron to the surface and this will eventually develop into spots of rust if they are not maintained (see Figure 4.4). This wall was given an initial wax coating but

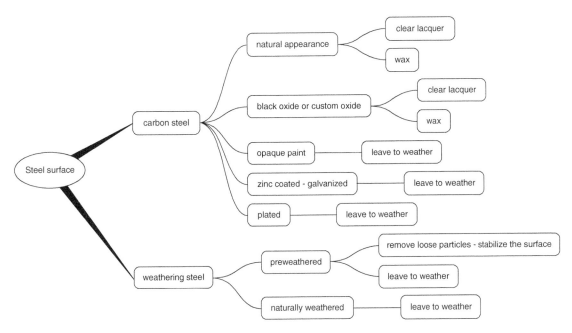

FIGURE 4.3   The various choices of steel surfaces.

was not maintained. After several years, the humidity formed condensation on the surface and created "islands" of rust. These can be removed and the surface protected. This is described in more detail in Chapter 8, but it is this natural tendency of the steel surface to react with the surrounding environment and the drive to return to the mineral form.

These blackened steels should be protected to inhibit the free flow of oxygen to the surface and to shed moisture. A light oil, clear wax, or clear lacquer coating would work effectively for interior applications if maintained. Exterior applications will face challenges from the changing environment, which will degrade these thin coatings. Exterior carbon steel surfaces will change more rapidly when exposed to the atmosphere and will require something more substantial such as a barrier of paint or zinc coatings (galvanized) to induce a cathodic protection to the base metal.

The use of weathering steels, either artificially preweathered or naturally weathered, will form a thin adherent oxide to provide long-term protection to the changes in the environment. The oxide, once formed, is inert and unconcerned with all but the most corrosive substances. For these steels, the reddish brown rust layer is the protective layer that slows down further decay of the base metal. Alloying elements within the steel and on the surface of the steel work to form an impervious barrier that limits diffusion of the iron and intrusion of moisture and oxygen.

When considering what to expect of a surface of steel, the environment it will experience will dictate how it will perform. The milder, interior exposure will be considered first. This environment is more stable, usually drier and reduced ultraviolet light radiation to degrade coatings. Interior

FIGURE 4.4   Rust on interior blackened steel wall.
Source: L. William Zahner.

exposures do have more potential human interaction and are more visually intimate and tactile than an exterior exposure.

INTERIOR EXPOSURES

Steels have, of late, moved from the strictly structural forms to decorative surfacing materials on the interior walls of high end restaurants, office lobbies, floors, and other interior spaces where the durability and tough look of steel is sought out. This rugged material can provide a pleasing impression, whether as a dark blue-gray tone or the dark reddish-brown character of weathering steel. The hot-dipped galvanized steel coating is another possible choice considered more and more as a finish surface alternative. The galvanized surface is the silver-gray spangled surface appearance produced by molten zinc that has solidified and bonded to the underlying steel. Plating is another option but is normally considered for smaller elements rather than larger, flat surfaces. Each of these will be examined in the following sections.

Carbon Steels

The carbon steels, when correctly sealed, will give the dark blue-gray tones. There are two basic surfaces, the cold-rolled steel surface and the hot-rolled steel surface. These are only applicable for interior surfaces. If used on the exterior, they would need to be coated with a durable paint over all edges, as well as the front and back surfaces. A challenge with using a clearcoat on an exterior application is the pretreatment that is necessary to enable the organic coating to bond. This can deter from the aesthetic originally considered.

As stated, the drive for oxygen is powerful with iron. On interior applications, if moisture is allowed to set on the surface, even the coated surface will lead to red rust spotting as weaknesses or porosity of the coating are uncovered. It will start as a fine haze that comes off when wiped. It forms from moisture condensing on the surface and reaching the steel through minute pores in the coating. When this occurs, the surface should be wiped down and waxed or recoated to seal the pores.

When the carbon steel has been oxidized to enhance the surface, a sound coating is still required. The process of darkening induces various compounds to react with the iron and form oxides or introduce a thin copper selenide coating or phosphate coating. These darkened films are very thin, only around 2.5 µm thick. They can slow the oxidation, but not prevent it entirely. A fine oil can be applied over these coatings to aid in shedding of moisture from the surface. Some of the coatings, phosphates in particular, are crystalline and have a slight roughness to them. These can be oiled with a light mineral oil. The coarse surface will absorb the oil, and this will both seal the surface from oxygen and act as a moisture repellant. You will need to maintain the surface by applying oil on a regular basis. Applying a good wax such as a microcrystalline or carnauba wax to the surface and buffing it in will create a temporary sound coating that provides protection and repels moisture.

The idea for any of these clear coatings is to allow the metallic color tone of the steel to show through and to not appear glossy and slick. Gloss flatteners are often incorporated into the coatings. These can give the impression that the steel is not coated. A thin layer of wax, once thoroughly buffed, will protect the steel without overtaking the natural appearance of the steel. Table 4.2 describes various coatings and the expected time between recoating. This assumes humidity levels below 50%. For every 10% increase in humidity, expect a 20% reduction in time.

The humidity will work to develop the red rust appearance on the surface, as indicated in Figure 4.4. Low humidity will extend the life between resealing the surface. As the humidity climbs, the metal stays wet on a microscopic level and any pores in the surface will eventually be breached. You can extend the times indefinitely by keeping the surface dry and maintaining a thin protective coat on the surface. This may involve removing the old wax and reapplying new wax. Removing the lacquer would be more difficult. It is suggested to place a thin, hard wax coating over the lacquer and removing and reapplying this wax coating rather than the more difficult lacquer.

There is something to be said for dual layers of protection. The wax coatings are hydrophobic when not deteriorated, hardened, and cracked. These coatings dry out and wear and need to be replaced. The sign of old wax is a yellowing or whitening. The acrylic coating offers protection from

TABLE 4.2   Performance expectations on interior steel surfaces.

Appearance	Coating	Time before slight rusting appears
Natural carbon steel – hot or cold rolled	Acrylic lacquer	48 mo
Carbon steel with darkening	Acrylic lacquer	60 mo
Natural carbon steel – hot or cold rolled	Wax	30 mo
Carbon steel with darkening	Wax	36 mo
Natural carbon steel – hot or cold rolled	Mineral oil	6 mo
Carbon steel with darkening	Mineral oil	9 mo

oxygen reaching the surface, and these are significantly harder than waxes, as well as resistant to chemicals.

When working with the steels on interior surfaces, the reverse side is also exposed to corrosion. If this surface is allowed to become wet, either from other moist materials in proximity or from condensation, it will develop rust that could potentially bleed out onto the edge.

Perforated steel surfaces demand special attention. The edges and both faces are subject to corrosion. Coating the perforated surface has to overcome thinning along the edge of the perforations. This can lead to corrosion forming along the edges of the perforations.

Cast steels will react similar to the wrought sheet or plate form. Same for pipes, tubes, rods, and bars. Moisture allowed to accumulate on the bare surfaces will cause corrosion to appear. Paint coatings are the normal means of protecting these.

If unprotected, interior applications are subject to humidity or moisture of any kind, they can be expected to develop iron oxide on the surface. These can be clear coated with an acrylic or other form of lacquer or they can be waxed. The same limitations are present with the relative humidity the object or surface is exposed to. See Figure 4.5.

The steel surface used in the interior applications, protected and kept dry, should last a considerable time. Columns, handrails, interior panels, and thin stamped ceiling panels have stood the test of time in many historic buildings. If unprotected, they will oxidize, and they can fingerprint. The fingerprints can lead to corrosion zones where the iron oxide will appear as reddish spots. These should be removed on a regular basis. Removal with a 99% isopropyl alcohol will not damage the clear coating. Light spray wax with lemon oil, rubbed into the surface, will help as well.

Steel window frames and steel door frames were a common feature in the late nineteenth century and early twentieth century. Even today these are considered for intriguing architectural projects. In 1912, a British firm, Crittal Manufacturing Company, Ltd., revolutionized the steel window with custom designs produced from hot-rolled steel shapes. Today they are called W20 profiles and have been further adapted for the more energy-efficient insulated glazing units. Similar

FIGURE 4.5   Steel surfaces on an interior.
Source: L. William Zahner.

units today are constructed from roll formed sheets of low-carbon steels. These would be coated with paint for long-term corrosion protection, often lead paint.

Today there are windows made from similar profiles but coated with galvanizing (zinc) or painted with zinc-rich primers.

It is remarkable how these steel window frames have survived in relatively good shape over all these years of exposure—exposures that were often industrial, urban regions. It is only where moisture has been trapped and held against the steel where indications of corrosion have appeared. Some of these steel window frames are approaching 100 years of service life.

Today modern steel window frames can accommodate different levels of insulated glass and different paint colors (Figure 4.6). The benefit of steel window frames is in the strength of the frames. Significantly smaller cross sections can be used to carry the loads compared to aluminum extruded window frames.

FIGURE 4.6   Modern steel window frames.
Source: L. William Zahner.

INTERIOR WEATHERING STEELS

Weathering steels will act just like the carbon steels on interior surface unless they have been preweathered. The surface of the weathering steels before they undergo weathering or accelerated weathering is similar in appearance to the low-carbon steels, either hot or cold rolled. The surface of this type of steel is indistinguishable from the carbon steels and performance will be similar to the carbon steels when used on interior surfaces unless it has first undergone a weathering process to build the oxide layer.

The weathering steels are generally provided in sheet or plate for art and architectural applications, however structural components such as angles, channels, tubes and beams are available

FIGURE 4.7   Weathering steel staircase.
Source: L. William Zahner.

in these alloys. These alloys fall under the classification of high-strength, low-alloy (HSLA) steels. They are carbon steels, nonetheless, but they have special alloying elements added to enhance the weathering ability of these steels. Chapter 3 discusses the preweathering process, but when using preweathered steel on interiors, added steps are required to discourage iron oxide particles from rubbing off on hands and clothing. Figure 4.7 shows a staircase made from preweathered steel.

Sealing the surface with a clear coating can freeze the loose particles onto the surface so it is important to remove as much of the loose particles as possible before coating; otherwise the coating will not effectively bond with the metal and can flake off. If there is too much fragmentation on the surface, the clear coating can come off with the oxide. Sealing the surface will also darken the appearance of the preweathered steel. Figure 4.8 shows a panel that is partially coated with a clear acrylic. There is a slight darkening of the surface where the clear coating has been applied.

Waxing, clear lacquers, or light mineral oils will darken the surface, turning the light reddish tones into darker, deeper reddish-brown tones. This is due to the metal absorbing the coatings into the porous surface and "smoothing out" the reflected light. When light hits the uncoated weathering steel, it is scattered from all the micro-surface irregularities. This scattered light delivers color to the

FIGURE 4.8   Preweathered steel coated with a clear coat next to an uncoated section.
Source: L. William Zahner.

viewer. When sealed, the light reflects from the sealed surface in a more consistent direction, less scattering occurs, and the surface looks darker, similar to when a stone or concrete surface is wetted.

Sealing the surface can be accomplished in a similar fashion to the carbon steels with a clear acrylic or a hard wax. The coarseness of the weathering steel surface aids in giving a "tooth" to bind the acrylic coating. However, the micro-coarseness will hamper the ability to work the wax into the weathered surface. Waxing can be difficult to polish out. This will make these coatings somewhat more difficult to remove in the event they need to be removed. Chapter 8 covers the event when an enamel as graffiti has been applied to the surface and the steps to remove this without damaging the metal surface.

Another process to arrive at a sound interior surface with preweathering steel is to work the surface sufficiently to remove all loose particles. This is done during the process of preweathering the metal. Lightly buffing the surface with a white Scotchbrite™ pad will remove the particles and leave a weathered steel appearance. It is critical to get the surface oxide stable and adherent. This will happen when the oxyhydroxide develops. Removing the loose surface oxide, which is the top layers of the more fragile ferric oxide, will leave the surface darker and render a depth of color into

FIGURE 4.9   Weathered steel appearance – interior surfaces.
Source: L. William Zahner.

the metal that will not rub off. It takes working the surface, almost burnishing the surface during the accelerated weathering process, to arrive at the levels where it will not rub off and the color is a rich reddish brown. Figure 4.9 shows images of 5 mm thick plate interior partitions.

Weathering steel used on interior surfaces must contend with the wear from human interaction. Instances when the surface is constantly rubbed or brushed against may end up having exposed base metal as the less flexible and more fragile oxide releases. Oils from handling a surface can enter deep into the pores of the metal and create dark tones, extremely difficult to remove. When constant wear is expected consider coating the surface with a clear urethane or acrylic paint. Also, a hard wax like carnauba works well. The wax can be removed and replaced more easily than the acrylic clear coating.

On interior applications of weathering steel, they should be expected to remain as initially installed. The color will not change. It may absorb moisture from the air but the surface color nor the surface texture will change in any appreciable manner.

EXTERIOR EXPOSURES

Steel surfaces exposed to the outdoor environment will change. Humidity, pollution, salts and sulfides, decaying organic matter, bird waste, moist soils, and condensation can create conditions that will alter the surface behavior of the steels.

For that reason, it is rarely the case where the low-carbon steels are left unprotected when exposed to the exterior environment. If they are not coated with enamels or galvanized with thick, zinc coatings, they will form the friable iron oxide and slowly erode away.

Clear coatings as protective barriers on steel exposed to the exterior environment can be challenging. Clear coatings, applied directly to the steel, do not perform as well due to the reduced surface preparation. The surface must be cleaned and degreased. The steel must be completely dry. Heating the surface before applying the clear coating will help remove traces of moisture.

Once coated with a clear protective film such as an acrylic or urethane, wax may be considered as a further barrier over the clear paint. In any event, maintenance must be performed on a regular basis to remove potentially damaging substances. Salts from deicing processes or seaside proximity will corrode the steel. A paint coating will need to be very impervious to withstand these exposures. Maintenance of the coating will need to be performed several times a year to have a chance to avoid corrosion.

Carbon steels are near universally coated with paint as protection. The steel provides form and strength and the paint is there for aesthetics and protection. Numerous everyday products are made of steel, lawn chairs, handrails, bicycle frames, hand tools, and the bodies of our automobiles are all made of steel generally coated with an opaque paint. We do not expect them to last indefinitely, or even a lifetime. These products will last only as long as the coating performs.

Steel is used extensively in structural applications and in artwork exposed to the outdoor environment. Figure 4.10 shows a Mark di Suvero sculpture made from structural sections, abrasive blasted and painted. With few exceptions, unless the artwork is made from weathering steel, the surfaces are coated with an organic or inorganic layer, usually several. For these to function properly, they need to be maintained periodically. Cleaning, ensuring no breaches in the paint at edges, welds, and bolts is important.

Carbon Steels

The attributes of steel—strength, durability, ease of welding, and cost—give the designer a material worthy of consideration. Unfortunately, there is one insidious impediment, the constant drive in

FIGURE 4.10 "Rumi." Designed by Mark di Suvero.
Source: Designed by Mark di Suvero.

iron's relentless desire for oxygen. This is a formidable constraint that limits the use of the carbon steels in art and architecture.

There are clear coatings that can be applied to carbon steels used on exterior elements, but these require constant maintenance to successfully protect the underlying steel. The ever-changing conditions of nature are relentless when it comes to ferrous parts. The clear coatings act solely as barriers to the environment. By their nature, they cannot impart cathodic protection the way zinc coatings can. As barriers, clear coatings are subject to ultraviolet radiation, under-film corrosion, adhesion issues, and porosity.

Copper selenide, blackening treatments, and other intermediate coatings provide some protection when they are kept oiled, waxed or clear coated. However, these coatings are very thin, and once the protective barrier wears away, you can expect these surfaces to corrode. Firearms and tooling are often coated with these darkening treatments, followed by a light oil. The oil acts as a hydrophobic barrier. Once the oil dries or wears away, the surface will corrode.

Wrought and Cast Irons

Both wrought and cast iron have been widely used over the decades as exterior ornamentation (Figure 4.11). Wrought iron has been the subject of great historic designs from Gaudi to the work of slaves who bought their freedom producing the beautiful ironwork that adorns New Orleans. Images of the wrought iron used in the past are depicted in Chapter 2, Figure 2.1.

Wrought iron is a metal that could be shaped and formed by heating, bending and hammering. It was so commonly used; it is often taken for granted. The Eiffel Tower was made from wrought iron, at the time called *puddled iron,* as was the framework used to support the copper skin on the Statue of Liberty; perhaps the last great architectural form made from this form of iron. Steel production processes along with the more manageable properties steel offered, overtook wrought iron as a major architectural form of iron.

FIGURE 4.11   Wrought iron.
Source: L. William Zahner.

Wrought iron has very low carbon, no more than 0.035%. When it is manufactured, it has a drawn out, fibrous strings of slag intermixed. This gives it strength and allows it to be shaped and hammered. The slag on the surface also aids in corrosion resistance.

It is difficult to weld in conventional methods, and the mechanical properties are highly variable, even in the same length of bar. Many of the older uses of wrought iron had bolted or riveted joints. Wrought iron is joined in the factory by heating and hammering under a powerful drop hammer while hot. The material flows together seamlessly when correctly performed.

Wrought iron used as architectural features has survived centuries (Figure 4.12). Its corrosion resistance is superior to mild steels and does not need to be galvanized or coated with zinc. It is normally finish painted for corrosion protect and aesthetic reasons. Today much of the wrought

FIGURE 4.12   Image of the wrought iron on Kansas City Museum.
Source: L. William Zahner.

iron is painted with a thick black or gray coating. Micaceous iron oxide was used to paint the Eiffel Tower and many other wrought and cast iron surfaces. It is a dark paint coating still in use today. Micaceous iron oxide, MIO, is made from a natural mineral ore refined to a composition of fine flakes of Fe_2O_3 crystals. It is set in a binder and applied over steel or iron surfaces to create a water resistant barrier.

Coatings on wrought steel usually fail around joints and laps. These areas are most susceptible to corrosion because they tend to hold moisture in areas where the coating is often thinnest or damaged from the process of joining. Keeping the seams and places where fasteners are used clean and protected is critical for the long-term success of wrought iron.

In recent times, many wrought iron forms are made from mild steel or cast steel. These do not have the same forming and corrosion resistant properties as the puddled or wrought iron. They are cheaper and more available. Mild steels can be welded using conventional welding processes and the mechanical properties are more consistent. However, the modern-day "wrought iron" are subject to faster corrosive activity and require constant attention to prevent corrosion. Figure 4.13 shows a decorative wrought steel form.

FIGURE 4.13   Wrought steel form made in the wrought iron genre.
Source: L. William Zahner.

Sources of wrought iron are limited. Most comes from recycled wrought iron that has been obtained from old structures and even mooring chains used centuries ago.

Cast Irons

The cast irons used in ornamentation is gray cast iron. Gray cast iron gets its name from the appearance of the inside of the casting when it has split or cracked. The gray color is from tiny graphite flakes interdispersed throughout the casting. See Figure 4.14.

Cast iron was used extensively in the late 1800s and early 1900s as a decorative and often functional form. Chapter 2, Figures 2.8, 2.9, and 2.10 show several examples of cast iron used as ornamentation and surfacing. Chapter 1, Figures 1.10 and 1.15 show the amazing cast iron work designed by the late Louis Sullivan.

Cast iron can be galvanized, but typically, it is not. It is usually coated in paint. Dark paints such as the micaceous iron oxides or combinations of zinc-rich primers and epoxies work well, but they tend to build up and conceal some of the beauty in the cast iron detail.

FIGURE 4.14   Cast iron detail. Painted with a thick protective coating.
Source: L. William Zahner.

CARBON STEEL AND STRUCTURAL STEEL

Steel structures, left unprotected, will corrode as thermodynamic conditions cause a gradual decline of the steel surface. It is a natural condition and underlies the reason iron is not found on the surface of the earth in the native form free of other compounds. This thermodynamic degradation of the steel surface happens quicker with the availability of moisture. Steel corrosion will be slower in arid climates but, over time the surface will show signs of decay as iron oxide forms on the surface as condensation collects and forms small corrosion cells. Figure 4.15 shows a painted pipe with corrosion along the edge where moisture would set and where the paint finish was mostly likely weakest.

Nearly half the steel that is produced today is destined for the building and construction industry. Structural steel beams, channels, angles, metal building panels, metal siding, metal decking, rebar in concrete all fall into this industry category. This category doesn't include appliances, ducting, and piping that also use steel and go into our buildings (see Figure 4.16).

FIGURE 4.15   Corrosion along the edge of a painted steel pipe form.
Source: L. William Zahner.

FIGURE 4.16   Share of steel production by industry.

The expectations of steel are strength, durability, and lasting, at least until the useful life of the structure is over. When the steel is exposed, engineers sometimes allow for a reduction of cross-section due to the potential of corrosion reducing the effective cross-sectional area.

To achieve long-term performance, the surface of steels used in art and architecture must be protected from corrosive substances and moisture. Adding some form of cathodic protection is common, usually in the form of sacrificial zinc.

Zinc-Rich Primers

There are several base coatings that are used on steel structures to provide a level of protection. The dark reddish color or dark gray color seen on heavy structural forms are zinc-rich primers that provide both barrier and galvanic protection to the steel surface. These zinc-rich primers are composed of zinc particles in an organic or inorganic carrier. These coatings can be coated further with a pigmented paint, providing a further barrier of protection. These coatings enable cathodic protection similar to the way galvanized steel protects steel; however, they lack the metallurgic bond and thus have a significantly shorter performance life.

These coatings are described as organic or inorganic coatings, depending on the nature of the carrier of the zinc particles. The organic primers are paint based (e.g., alkyds, epoxies, and urethanes) and will often have a pigment of gray or red. They have various levels of zinc particles suspended in the coatings and are classed in different concentrations from 65% zinc particles to more than 85% zinc particles. The organic coatings are easier to apply in the field to assembled and welded surfaces. The inorganic coatings are silicon-based carriers and offer slightly better galvanic protection and can have a greater concentration of zinc. These inorganic coatings are not easily applied in the field and do not take a top coating of paint well. The steel must be prepared, usually by thorough abrasive blasting of the surface, before application.

For both coating types, organic or inorganic zinc prime coatings, thorough preparation of the surface is required. The steel must be clean and free of oils, grease, and oxides, and the normal mode of preparing the steel is to abrasive blast the surface. In North America, the standards for preparing steel are establish by "The Society for Protective Coatings" or SSPC along with the National Association of Corrosion Engineers, or NACE International.

The preparation of the steel surface for these protective coatings is paramount for long-term success. For the organic coatings, generally the specification is for an SSPC-SP6/NACE 3 Commercial Blast Cleaning. This allows for some level of oxide to remain. Streaks and mottling due to partial abrasive blasting can be visible, but no more than a third of an area. In practice, this can be somewhat subjective. You want the entire surface covered and all loose substance removed, as well as any oils. Otherwise, the coating will not adhere well in these areas (see Figure 4.17).

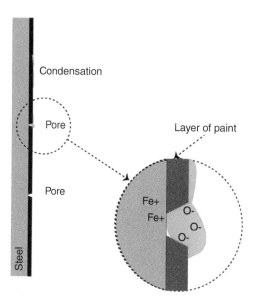

FIGURE 4.17   Pore in paint coating.

FIGURE 4.18   Near white blast of a steel plate.
Source: L. William Zahner.

For the inorganic coatings, the surface must be given a SSPC-SP10/NACE 2 near-white blast cleaning treatment (Figure 4.18). This is a very thorough cleaning of the surface, and no more than 5% of the surface can show any form of stain or discoloration.

The abrasive blast treatment prepares the metal surface and gives a bit of tooth to enable the coating to key into the surface. For a list of the SSPC designations, please refer to Appendix D.

When applied correctly, these coatings can be expected to provide several years of service before signs of iron oxide appear. Salt-laden environments, both coastal and deicing, can degrade these coatings in less that 18 months, unless they are maintained or further coated with another barrier coating.

Often, these coatings are used to touch up welded or damaged sections of galvanized steel fabrications. From an aesthetic perspective, this is a mistake. The paint surface will oxidize at different rates and show as spotting on the galvanized surface. The zinc-rich coatings may initially look like the galvanized, but they will appear very different in gloss in a very short time.

Another silver-gray coating that is often used is an aluminum paint. Similar to the zinc-rich gray paint, this contains aluminum. It goes on brighter and more reflective. This can be considered when touching up new galvanized steel. It stays bright similar to the unweathered zinc on the galvanized surface.

Galvanized Steel

Exterior uses of steel with the galvanized coating is common for architecture and art when the steel will be subjected to weathering. It is an excellent way of achieving an aesthetic along with strength.

When large architectural parts are hot dipped the finish will last and protect the steel. You should avoid post welding or field welding. Once hot dipped, the connections should be bolted. The holes should be preset to allow for the inner edges of the holes to be coated.

Galvanizing of steel has been in practice for nearly 300 years. First developed by the French chemist, Paul Jacques Malouin, back in 1742 when he presented a paper on dipping iron plates into molten zinc. Later, near the end of the century Alessandro Volta with influence from Luigi Galvani and so named for Galvani, wrote on the principle of cathodic protection of iron. The process was later patented by Stanislas Sorel in 1837. He called his "new process" galvanizing and set factories up around Great Britain and Europe to coat iron objects in zinc. This is also the beginnings of an intriguing casting method called "slush" casting and will be discussed in more detail in the book on zinc.

When steel-rolling mills came into widespread use in the late nineteenth century, the process of galvanizing these sheets for corrosion inhibition advanced. Today it is common practice to coat sheet steel in zinc using the method. Even when paint is to be applied, pre-galvanizing the steel is the normal process.

For large steel structural forms and heavy steel products, zinc coatings via the hot-dipping method can produce a long-lasting, corrosion-resistant coating that will stand against the environmental exposures and keep the steel corrosion free. The coatings applied on these are usually in the range of 0.5–0.10 mm thickness. Galvanizing protects the steel by supplying the electrons and forming a layer of zinc oxide and hydroxide. This is a form of cathodic protection where the zinc sacrifices itself to protect the steel. Thick zinc coatings applied by galvanizing will protect the underlying steel even when scratched through to the steel. The cathodic protection of the zinc will extend over to the exposed metal as the zinc acts as an anode when an electrolyte (moisture) is present.

Coating steel with molten zinc, when performed correctly, forms a metallurgic bond between the two metals. Unlike paint or oils, the zinc and steel actually interface with one another as iron diffuses into the zinc layer. See Figure 4.19. The zinc will not peel off, crack, or chip from the steel because it is now part of the surface. Some iron diffuses into the zinc layer turning it into more of a continuum as the pure zinc layer on top turns into a mixture of steel and zinc at the interface.

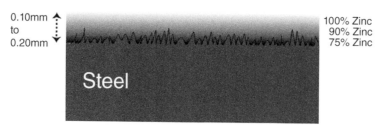

FIGURE 4.19   Cross section of hot-dipped galvanized on steel.

When specifying galvanized steel for art and architecture, it is critical to work closely with the galvanizing facility to ensure they are aware of the end use. Galvanizing is an industrial process and the handling does not always meet with minimum artistic or architectural compliance. The newly galvanized steel parts can be stacked on one another, inducing surface scratches. This may not necessarily harm the corrosion behavior because of the way the zinc offers sacrificial protection, but it will impose a visual shoddiness. Other issues involving poor handling might be footprints or dirt on the surfaces. There may be contamination in the galvanize that produces clumps and roughened spots. Perforated steel surfaces can have strings passing through and across the holes.

Industrial processes often have different objectives of speed and output. The workforce is used to handling large sections into and out of baths of acid and molten steel. They are used to moving things in quickly and out. Asking them to implement specific care and handling will interfere with the cost model and the typical operating behavior.

This condition is true with steel in general. Most steel facilities are rarely involved with the creation of aesthetic forms. Wrought cold-rolled sheet is the exception, since much of this is destined for secondary facilities involved with stamping, forming, and bending operations and they will want clean blanks free of scratches and excessive oxide. Hot-rolled plate, bar, tube, and pipe are rarely handled with art and architecture as the end product in mind. The importance of monitoring the supply chain from a surface quality context will be necessary to achieve an acceptable result. The supply facility or the galvanizing facility may require more financing to cover additional costs, but it is wise to get them involved in the understanding of the end goal.

Different forms can be galvanized. Figure 4.20 shows, in the upper image, corrugated panels made from galvanized steel coil while the lower image shows steel tubes, welded, and hot dipped as assemblies.

WEATHERING STEEL

The most common exterior steel used in art and architecture is weathering steel. Weathering steel is an HSLA alloy. A corrosion-resistant, natural surface beauty few other materials can achieve will develop over time and exposure to the environment. Or, more often the case today, an accelerated oxidation by preweathering the steel in a controlled environment will develop this oxide layer in advance.

Weathering steel is a specially designed steel. It is not simply carbon steel but a carbon steel with alloying elements that develop into a highly corrosion resistant oxide. This oxide develops in layers as moisture and oxygen react with the iron and the other elements on the steels surface. See Figure 4.21.

Natural weathering takes time. As the metal is first exposed to the environment a thin oxyhydroxide forms. Usually accompanying this oxide are other oxides of steel, ferric oxide, and ferrous oxide. The oxide forms into large crystalline peaks separated by areas of no oxide. This is not unlike

FIGURE 4.20   Galvanized steel in sheet and fabricated forms.
Source: L. William Zahner.

regular carbon steel. Electrolyte on the surface generates zones of localized anodic and cathodic electrical potential due to variations on the surface. With weathering steel, these are enhanced by the occurrence of copper ions on the surface. The initial surface is a streaky, uneven oxidation, as moisture from rains or condensation create minute cells of electrochemical polarity. Figure 4.22 shows a newly installed wall and the same wall after about one month. The oxide is very new and weak.

These large crystalline peaks are weak and will easily be knocked over when brushed. Some of these regions, those of ferrous oxide, will dissolve in water and redeposit on surrounding substances. This is the characteristic stain associated with many weathering steel surfaces. As more moisture arrives on the surface, the oxide thickens and covers more of the surface. At the same time, some of this semi-soluble material comes off the surface and deposits on regions below the metal. This will

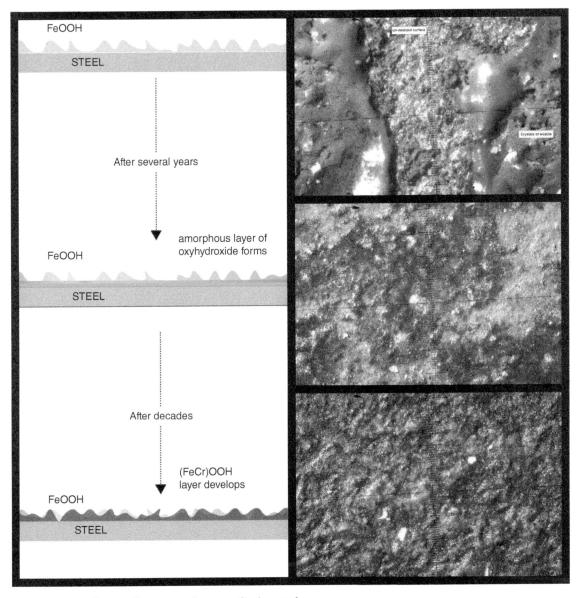

FIGURE 4.21   Stages of layer growth on weathering steel.
Source: L. William Zahner.

FIGURE 4.22   Initial rust.
Source: L. William Zahner.

go on for several months before the iron oxide thickens to the point where it no longer comes off. Figure 4.23 shows some light bollards on a concrete surface. The staining occurred initially after a few months and has now stopped.

Many designs incorporate drip edges to redirect the stain or rock drainage to allow the stain to be captured and not run onto other substances. The amount of staining can be significant and needs to be planned for. Figure 4.24 shows a couple of approaches used to capture the runoff stain.

Through exposure to humidity, sunlight, and time, the oxide thickens, and diffusion from the steel alloy outward coupled with the thickening layer of oxide on the surface forms another intermediate layer, one lacking a crystalline surface. This is the layer that forms the impervious, corrosion-inhibiting barrier. The color deepens from an orange-red to a reddish brown. It may have islands of ferrous oxide still forming on the surface. Figure 4.25, image 1 in the upper left, shows the "islands" of ferrous oxide that can add to the stain. This layer continues to develop with

FIGURE 4.23   Staining.
Source: L. William Zahner.

mixtures of iron (II) hydroxide and iron (II,III) oxide forming and thickening as seen in image 2 on the upper right. This takes anywhere from 12 weeks to 6 months before it thickens and tightens to the oxide shown in image 3. This is a mixture of the hydrated iron oxyhydroxide and iron (II,III) oxide.

As more time passes, the intermediate layer develops into a layer of iron and chromium oxy-hydroxide not unlike what is seen in the lower right, image 4. The oxide thickens as the iron (III) oxide with the red-brown color forms along with the iron oxyhydroxide. In a mineral sense, this layer is a chromium-enriched Goethite. It is this layer that forms into a hard crust that is both abrasion resistant and corrosion resistant. The color turns into a reddish-purplish brown. No staining occurs, no ferrous oxide develops. The surface reaches a point of equilibrium that should perform well into decades ahead. A state where the reddish color goes deeper to a purplish red, some describe as the color of eggplant. Figure 4.26 shows a chevron roof made of 2 mm thick, weathering steel. This surface is over 50 years old.

When the project requirements are to have the steel weather naturally in the open environment, beginning with clean, oil free and scale free surfaces is critical. Oils on the surface will impede the development of the oxide. Until the oil is removed, either through weathering or cleaning, the

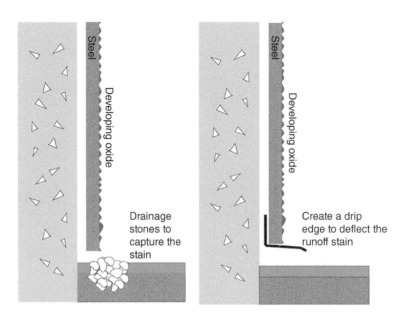

FIGURE 4.24 Runoff stain deterrent.

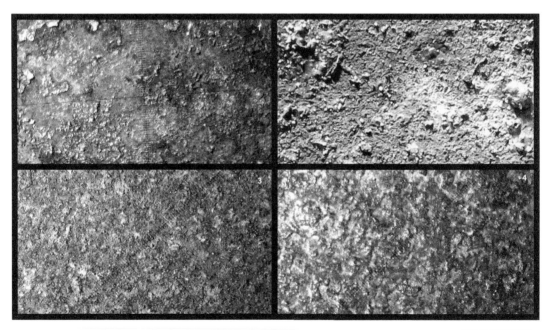

FIGURE 4.25 Weathering steel microscopic layer development.
Source: Courtesy of Solanum.

FIGURE 4.26   Chevron roof of weathering steel. Over 50 years of exposure.
Source: L. William Zahner.

desired oxidation will not develop. Removing all oils and starting with a clean dry steel surface is necessary. If there are small scratches from handling or processing, these will eventually be concealed by the development of the oxide. Large scratches, dents, and mars will remain. Even rust from banding or oxidation created from moist wood blocking will remain on the surface for a long period of time. It may be necessary to abrasive blast the surface to remove these marks or replace the steel with clean unmarked steel. Otherwise, you will need to accept these aesthetic alterations.

The protective oxidation will not grow the same on scale that is present on the surface. Scale is a hard, glass-like substance. It contains iron oxides and other substances that formed on the surface of the hot rolled steel. Cold-rolled and hot-rolled and pickled steel is usually free of scale. Thick, hot-rolled plate can have remnants of thick scale on the surface. Oxidation will be different where scale is present. Scale can be removed by abrasive blast. There will be a roughened surface remaining where the scale once was. Like an eroded scare, the area left after the scale is removed is of different texture and at a shallower plane.

TABLE 4.3   Natural weathering vs. Preweathering.

Condition	Natural weathering	Accelerated weathering
Initial color	Consistent-streaky	Variable from element to element Reddish brown
Color after 1 yr	Consistent orange	Variable, but reduced
Color after 5 yr	Consistent reddish brown	Consistent deep brown
Staining	Significant for first years	Significant reduction in staining from the outset
Streaking	Significant for the first year	Little if at all
Rubbing off	Significant for the first year	Little if at all
Flaking of surface	Chloride exposure can cause some to flake	Should be minimal if at all
Cost	Low	Moderate

Additionally, any markings such as from grease pencils and marking pens will show as these will not weather over. They should be removed prior to installation. Adhesive from tapes, oils from workers hands, or equipment will also lead to an initial corruption of the oxide layer.

The color of weathering steel allowed to develop naturally is more even and consistent than the initial preweathering steel. HSLA steel allowed to develop the oxide naturally develops the surface oxide and the color tones all at once in a consistent atmosphere. It is not to say that preweathering is not capable of this, it is simply the variables the steel is subjected to, take time and small variations in the weathering chamber over time can lead to slight color variations. See Table 4.3.

The decision to allow a surface made of HSLA, weathering steels to oxidize naturally or to preweather the surface to accelerate the oxidation is an aesthetic one. The natural weathering will take time, several months with average rains and level of humidity that form condensation on the surface. Natural weathering will take significantly longer in dry, low-humidity regions. Staining of adjacent surfaces as the oxidation grows is an issue that should be considered and addressed. Initially, for the first several months, the surface is very streaky and rust stains are produced over surfaces below the oxidizing metal. The rusty stain can be orange to reddish orange. It is very adherent and difficult to remove.

Preweathering the surface can add cost and time to prepare the metal, but the stains are avoided as the soluble rust particles are contained in the weathering chamber used to create the oxide coating. Figure 4.27 shows the controlled preweathering of steel panels.

FIGURE 4.27   Preweathering large steel plates.
Source: Courtesy of Solanum.

PREWEATHERED STEEL

It is more and more common to have the weathering steel undergo a process whereby the protective oxide is artificially generated. Chapter 3 discusses more in-depth on how this is achieved. The goal is to advance the development of the oxide to a point where the soft, friable crystals that first grow on the surface are replaced by a denser and more adherent oxide layer. A layer that does not easily wash off during subsequent rain events. The benefit of this is several fold. By preweathering the surface, you take control of the important initial development of the oxide. Contaminants such as mill oils are removed, scale is removed, and the surface begins its chemical reaction with a more consistent start. Preweathering establishes the direction of the oxide growth and allows for the development of the primary oxide on soffit surfaces, hidden surfaces and other regions that if set to oxidize naturally may be delayed or develop a different appearance from other more exposed regions. Figure 4.28 shows a preweathered surface with a soffit of similar color and tone to the vertical region.

FIGURE 4.28   Preweathered surface on a vertical wall and soffit return.
Source: L. William Zahner.

This process of preweathering takes time and must be undertaken in a controlled environment to ensure success. Keep in mind that this is a chemical reaction on the metal surface. Temperature, humidity, chemical composition, electrolyte, and time all play a significant role in determining the efficacy of the reaction. The reactions are slow to occur when the temperature is low or when there is a lack of surface moisture to initiate the reactivity.

The process of preweathering is dependent on both temperature and moisture applied in a controlled manner. A cycling technique of wet and dry periods repeated while in a temperate relatively constant environment is needed to accelerate the oxide layer.

Signs of failure occur when the oxide layer lifts from the steel itself. This occurs when the oxide grows rapidly and dries out. It will begin lifting from the surface as the friable crystals form too quickly and separate from the underlying steel layer. The surface may appear to meet color tone levels that indicate a rich oxide growth but, once it dries out, it begins to flake off the surface.

When correctly formed, the oxide is unbroken across the surface and there is no underlying steel exposed. When this is achieved, staining of adjacent surfaces is minimal. Most of the loose particles are removed in the preweathering chamber, and what remains is a very adherent oxide surface.

This initial stage sets the pace of oxidation by making the surface of the steel active. When preweathering, you first need to establish numerous tiny corrosion cells on the surface. It is the timing and intensity of these corrosion cells that is critical. Too aggressive, and the initial oxide grows away from the base metal. Too mild, and the surface will obtain an oxide that will actually slow the process down by forming a barrier over the surface and preventing oxygen from reaching the iron.

At this point, the preweathered steel could be set to naturally weather further. However, much of this oxide will sluff off the surface and stain adjacent materials around the steel. On some projects, this may not be an issue. Keep in mind the surface is very delicate and handling can smear the long crystals that have formed on the surface.

It is recommended that the oxide be further built out before handling to thicken and bond the oxide to the base metal. Figure 4.29 shows a preweathered wall surface composed of overlapping shingles. The wall has been exposed for several months.

As the oxide grows, gaps on the surface begin to fill in. The color darkens and the underlying intermediate oxide begins to develop. At this stage, the surface is still prone to smear, but there is an improved density of the oxide. Figure 4.29 shows a surface at the intermediate level. Stability of the surface is achieved and more cycles of wet and dry from natural effects will fill in the gaps and thicken the oxide. Staining will still occur as microscopic gaps of steel form the more friable ferrous oxide and ferrous hydroxide. Some of this will still dissolve in water and redeposit on adjacent surfaces. Controlling where moisture will drain from the surface is critical. Figure 4.30 shows weathering steel plates offset from a white wall but at the joint between panels moisture collects and stains the surface.

FIGURE 4.29   Shingles of preweathered steel.
Source: Courtesy of Solanum Steel.

FIGURE 4.30   Staining at joint between weathering steel plates.
Source: L. William Zahner.

Further cycling of wet and dry cycles in the controlled environment will develop the deep oxide rich in oxyhydroxide on the surface. The color darkens considerably. Gaps in the oxide layer are filled in and the formation of the intermediate, amorphous oxide has begun. Staining is reduced considerably. The surface is hard, and rubbing the surface does not leave corrosion products on the hand or a white cloth. See Figure 4.31.

Once this level is achieved, the surface is very stable and is similar to what would occur if the weathering steel was allowed to weather naturally for 10–15 years. The surface will continue to deepen on exposure to the atmosphere, but the rate will be very slow.

Preweathering perforated or cut HSLA steel is difficult. Attempting to preweather the cut edge to the levels of the surrounding surface is not easily achieved. The metal is homogenous throughout. The same interstitial elements of copper, chromium, and phosphorous are on the surface of these cut edges but methods used to cut the openings and the preweathering process will have an effect on the oxidation growth.

For instance, if the piercings or perforations are created by plasma or laser, there is a slight morphology along the edge as the steel was heated to melting. The oxide that reforms at the cut

FIGURE 4.31 Preweathered steel signage. Note reduced staining.
Source: Courtesy of Solanum Steel.

edge are of a different character than the balance of the surface. If the perforations were cut by waterjet, there may be remnants of the garnet abrasive used in the water.

These oxides and abrasives may slow the oxidation down, but it is the physics involved with wetting and drying a ledge or return at the cut edge that plays the greatest role. It is difficult to wet the upper return on an opening as well as the vertical returns. The lower region may stay wet longer as moisture sets on this small ledge.

In either the natural weathering or preweathering, this can lead to streaking of the surface as oxide particles form and slowly leach out. Refer to Figure 4.32.

Eventually the edge will develop the thicker oxyhydroxide but until this occurs you can expect a brighter-colored stain on the darker background. The only way to lesson this staining is to install the works and wet the surface frequently allowing it to dry between wet cycles. If the surface is visible from both the face and reverse sides, there can be streaking on either side or both. More time and exposure are needed to overcome the thinner oxide that first develops around the edges of perforations.

FIGURE 4.32   Streaking of perf edges.
Source: L. William Zahner.

MARINE ENVIRONMENTS

The use of weathering steels in marine environments has been analyzed. An eight-year study[1] found the weathering steel exposed to a marine environment formed a corrosion product that rapidly grows to a limiting thickness and from there very little change. The study found the corrosion rate of weathering steel fell dramatically once the oxide developed on the surface. From the study they extrapolated the time it would take for weathering steel to lose 0.25 mm (0.01 in.) in thickness would be 34.8 years (Figure 4.33).

Figure 4.34 is a beautiful example of the use of weathering steel near the coast. This example is off the coast of British Columbia, 10 m from the ocean. The weathering steel was preweathered in a controlled environment then installed onto the structure. After six years of exposure, the surface was found to be virtually unchanged. Measurements taken of the plates show the metal is actually

[1]H. Townsend and J. Zoccola, "Eight-Year Atmospheric Corrosion Performance of Weathering Steel in Industrial, Rural, and Marine Environments," in Atmospheric Corrosion of Metals, ed. S. Dean and E. Rhea (West Conshohocken, PA: ASTM International, 1982), 45-59. https://doi.org/10.1520/STP33185S.

FIGURE 4.33 Amherst College, Science Center designed by Payette.
Source: Images courtesy of Jeff Abramson.

thicker due to the oxide growth on the surface. The oxide will have a slightly greater volume than the base metal. The oxide would not rub off the surface and the surface appeared to be performing very well.

The sea coast exposure creates a rougher surface as the metal weathers. Figure 4.35 is a close up of the surface of the weathering steel plates used on the residence. The coarseness adds to the color. The thick oxide does not rub off or bleed onto adjoining surfaces.

Chlorides makes for a much more thermodynamically active surface and there is some potential of sluffing off of some of the preestablished oxide. This also happens where deicing salts are prevalent and are allowed to remain on the surface. Deicing salts contain chlorides as well.

Weathering steel will perform well in marine environments as long as the surfaces are well drained. The oxide will thicken and resist corrosion as long as it is able to dry. You should expect some streaking at joints and drainage collection zones.

FIGURE 4.34   Pender Island Private Residence. Designed by Tony Robins.
Source: By Ema Peters.

FIGURE 4.35   Close-up of the surface.
Source: L. William Zahner.

URBAN POLLUTION

Preweathering steel surfaces will perform well in heavily polluted, urban environments. The surface becomes resistant to nitrates, sulfides, and phosphates as the copper and nickel alloying elements combine to form an insoluble salts that tend to fill the pores in the oxide. Sulfur dioxide, the damaging substance in acid rain, will form sulphate salts with nickel and copper on the surface.[2] These are insoluble salts and combined with the oxide create a formidable barrier against corrosion.

Figure 4.36 shows preweathered steel panels used on a façade in Jakarta, Indonesia. The panels are 3 mm thick, solanum™ preweathered steel. These are performing very well in this humid, urban environment.

Good Design

If the importance of the metal surface is not transferred to the workers handling and installing this special steel, they will often inadvertently damage the surface. Many workers do not appreciate that a steel, particularly a rusty steel, could actually be the finish surface. Protective films or coverings will not stick to the surfaces of preweathered steel, so they need to be packaged in ways that the

FIGURE 4.36   Preweathered steel in a humid, urban environment.
Source: Courtesy of Solanum Steel.

[2]Chawla, S.L., Gupta, R.K., *Material Selection for Corrosion Control*, ASM International, p. 112.

FIGURE 4.37   Open shipping crate with separators to keep the plates from rubbing.
Source: L. William Zahner.

surfaces do not rub against one another during handling and transit. Figure 4.37 shows a stack of custom-perforated plates in a steel crate with separators between the plates to keep them from rubbing and keep them dry.

The thought might be that the surface is indestructible or the surface is intended for outdoors, so why not simply store it outdoors? For most metals, allowing water to rest on the surface often leads to problems with the finish. Weathering steel is no exception. Weathering steel should be allowed to drain well if stored out of doors. Steel surfaces placed face-to-face will lead to significant corrosion on the surface.

Figure 4.38 shows a surface of preweathered steel plates destined for a stadium in the Middle East. The panels were made from 7 mm thick weathering steel plates. They were preweathered

FIGURE 4.38   Weathering steel plates for al-Basrah Stadium.
Source: L. William Zahner.

and specially packaged so one plate would not rub against the other. Yet all these precautions were ignored. The workers sorted the panels out onto the desert floor and walked on them. Even the inspector stood on the metal plates while still packaged.

On this same project, once the panels were installed, the roof drainage system was temporarily directed over the surface. This created streaks and stains as the preweathered surface was confronted with moisture and debris from the roof.

Allowing water to set on the surface will create a stain as some of the iron ions diffuse into the water and oxidize. This creates a lighter-color stain that will remain on the surface for years.

FIGURE 4.39   The al-Basrah Stadium designed by 360 Architecture.
Source: designed by 360 Architecture.

The project was in an area of the world where weathering steel is not common, however. Its adamantine nature was taken to task by the way the material was handled. In the end, for the most part, the stadium is an attractive addition to Iraq. See Figure 4.39.

It is crucial that people at all levels, from sourcing the metal, processing the metal, delivery, and installation, understand the delicate nature of the initial oxide surface. It can be abstruse to some because if falls into an area of construction that appears, at first blush, rugged and inconsequential. Getting the correct results demands a minimum level of care.

Preweathered Steel Requirements

- Supply chain must acquire the correct alloy of steel.
- Supply chain must get steel that is clean and free of scale.

- The surface of the metal must be free of oil and grease.
- The surface must have all oxides removed by abrasive blast treatment.
- Immediately following oxide removal, treat with appropriate oxidizing agent.
- Properly handle and dispose of all hazardous materials.
- Follow with water rinse preferably deionized water initially.
- Allow to dry.
- Mist surface in controlled environment:
  - Temperature above 20°C
  - Relative humidity above 60%
- Allow to dry.
- Repeat mist and dry cycle until color and surface stability is achieved.
- Store in a clean, dry area until ready to ship.
- Package so that surfaces do not rub against each other during shipping.
- Store in a clean, dry area at project until installed or allow ventilation and drainage.
- Do not allow water to rest on the surface.

CHAPTER 5

Designing with the Available Forms

We fail to recognize situations outside of the context we learn them in.

- Domain dependence

Source: Modified from N. N. Taleb, Domain dependence

INTRODUCTION

Steel and iron have been used in various forms for centuries. But, unlike other metals, it did not begin as a cast form. Our ancestors did not have the means to achieve the high temperatures needed to melt iron. They could soften it using the energy sources of the time but, making iron fluid was not initially possible. Iron began its relationship with mankind more modestly than what we experience today.

The manufacture and forming of steel require energy. A lot of energy. When steel is manufactured from ore and shaped into plates, sheets, and I-beams, energy is needed. To overcome the strength of steel often induction heating is required to shape and form heavy sections. Figure 5.1 shows a thick steel plate that has been heated to red hot in order to allow curving into a tight cylinder.

Steels, the name given to the family of the most common alloys of iron, are available in various forms to be further shaped by industry. These forms are used by the fabricator of art and architectural products as well as engineered products and structural shapes. In art and architecture, steel forms used to fabricate the surfaces and objects of art, fall into one of two categories, cast or wrought.

Steel castings are less common in the art and architecture environments because steel is more difficult to cast – more energy is needed and because they are subject to corrosion – the release of this stored energy. Still, castings of steel are seeing a resurgence due to the low cost of the base metal and the strength it affords.

163

FIGURE 5.1   Steel plate shaping.
Source: Courtesy of Taylor Forge Engineering Systems.

Cast

Sand castings

Permanent mold casting

Investment casting (lost wax)

Ceramic mold

Wrought steels are more common in art and architecture. Particularly with the weathering steel sheet and plate and coated steel shapes and forms. Steel presents the designer with a material that is both economical and strong.

Attributes of Wrought Steel

- Corrosion resistant (weathering steel)
- Good strength
- Durable

- Holds an edge
- Magnetic
- Hot working ability
- Hardness increased by quenching in water
- Could function as molds for other metals
- Malleable
- Custom finishing
- Recycle ability
- Low cost
- Color – earth tones (weathering steel)
- Color – steel gray

EARLY FORMS OF IRONWORK

Our ancestors did not have the energy sources to work with iron the way we have available today. Iron and its alloy steel did not arrive in common use until the later part of the nineteenth century. Armor, knifes, swords, and firearms were manufactured from wrought iron and eventually cast iron in the fourteenth century. The available iron for industry was limited. Figure 5.2 shows a cast iron cannon. Cast iron was stronger than cast bronze.

FIGURE 5.2   Cast iron cannon.
Source: L. William Zahner.

Since iron is not found naturally in its pure state, the source of this early iron came from meteorites. Meteoric metal is iron, rich with nickel. The early use of iron most likely involved cutting chucks of iron from the meteorite with stone tools as the Eskimo's were found to have done. They would remove chunks of the meteorite and put them in walrus bones to use as cutting tools.[1]

In the Anatolia region, rich deposits of magnetite were used by the Chalybes to create iron by roasting the magnetite in layers of charcoal. A mass of spongy iron would result. This mass could be hammered with wood and stone into shapes such as axe blades or sword blades. This early iron was a form of wrought iron. Ductile and malleable, this form of iron could be hammered and shaped into simple forms, then hardened when quenched in water.

WROUGHT IRON

Wrought iron was made by combining iron with slag from the furnace. This slag was waste created during the melting of ore and when combined with the molten iron to about 2%, a fibrous mass would form. This fibrous mass, once solidified, could be hammered out to knock away excess slag. The fibrous wrought iron remained and could be shaped and flattened.

It would take about 100 kg of charcoal to make 1 kg of wrought iron. It was a very drawn-out, labor-intensive process, but the result would be a very corrosion-resistant form of iron. Early wrought iron could provide a useful life for hundreds of years before it would eventually be consumed by corrosion. Figure 5.3 shows wrought iron steel bars used in an old prison in Italy.

FIGURE 5.3   Wrought iron bars.
Source: L. William Zahner.

[1]Delmonte, J., (1985); Ancient Metallurgical Practices, Origins of Materials and Processes, Technomic Publishing Company, Inc.; Lancaster, PA. 238.

The time-consuming and slow process of producing the wrought iron made the metal rare and valuable. Products centered on tools and weaponry. The process of making wrought iron advanced in the 1700s and 1800s. The Aston Byers process was developed in the 1700s. Further advances to the manufacture of wrought iron during the early 1800s made wrought iron in larger quantities and enabled the Industrial Age. This process used pig iron, with a very low carbon content. The pig iron would be mixed with around 2% slag, which would form inclusions during solidification. Further shaping of the iron and slag would produce the fibrous form we associate with wrought iron.

Wrought iron was the main form of iron for centuries. The Eiffel Tower was made of wrought iron. Wrought iron processes are nearly as old as casting but commercial uses have for the most part changed to mild steel in the wrought form. Wrought iron has all but disappeared and is only available from small foundries, where it is either manufactured using the old methodology or recycled.

WROUGHT FORMS OF STEEL

Most forms of steel used in art and architectural today are in the wrought forms. Not the wrought iron of old, but the forms of sheet, plate, wire, rods, and shapes. The word *wrought* means "to work." Essentially, wrought forms of metals are those that have been first cast then worked into forms with a linear nature. Common with the wrought forms is a direction to the grains as the metal form is rolled out in a linear direction, the grains are stretched in this direction and this will have an effect on the mechanical properties. This is what gives most wrought forms an anisotropy behavior. Properties of strength are slightly different in one direction versus another. Other physical properties such as thermal expansion, texture, and even, to a degree, corrosion resistance have a directionality characteristic. This is due to the process of manufacturing the form. The microstructure that makes up the steel has a linear orientation.

The process for creating the various forms begins the same but then branches off into production sequences specific to that form. That is, the metal begins in a molten stage then it is cast and rolled in a specific mill station for that form. There are stations that produce the hot-rolled plate, and there are mill stations that produce the wire form, essentially each form of steel has its own particular path that imparts characteristics to the metal. When a station begins to make wire, the initial cast metal billet used to make the wire is dedicated to only wire. The forms all have various limitations due to the initial constraints of production and subsequent constraints imposed by the manufacturing processes. Figure 5.4 shows the paths taken to create the various forms of wrought steel.

The molten steel is first destined for continuous casting or ingot casting, depending on whether the steel Mill is a specialized mini-mill or large integrated mill. If the steel is to be processed in the continuous casting method, the bloom, billet or slab is the product created in the cast. The specific cast product is subsequently heated and delivered to the specialized mill for creating the initial shapes for industries, such as plate, wire and structural steel forms.

Once the path is chosen, there is no stopping and redirecting. Billets that are destined for the wire mill are completely consumed in making the wire. Same for the other forms steel takes.

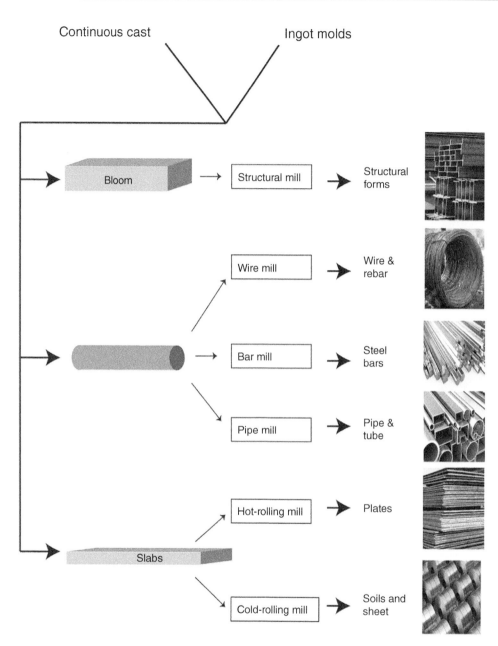

FIGURE 5.4   Paths of the initial fabrication of steel forms.

These various forms are produced at the large Mill or Mini-mill for industry needs. The alloying, mechanical properties, temper, annealing, pickling, and oiling are performed at the Mill source to meet the customer requirements. Many of those customers are the large warehousing and secondary processing facilities that supply smaller, more diverse industries. The form, alloy and heat treatments are determined by market demands as well as specific orders from large manufacturers such as the automotive industry and building industry.

Small orders are subject to availability. Many of the warehouses stock a diverse array of products in stock lengths. Some will process the material further by adding value operations of cutting, slitting, even finishing. But the small order must often adapt to what is available and this may require purchasing larger quantities than what is needed from different parts of the country or world. For example, steel pipe is used extensively in the oil and gas industry. Warehousing of pipe will often be in these regions and the pipe or tube will be in dimensions used extensively by oil and gas operations. It is important to work with the distributor and warehouse to determine the most optimum and available size and dimension that will meet the design needs.

There are numerous forms of wrought steel. Wrought steel is used to make various shapes destined for artistic and architectural designs. The availability of the basic forms of carbon steel and the economy provide a ready material for manufacturing into useful items. Figure 5.5 shows images of traditional wrought steel handrail made from bars and tubes.

The wrought forms are those forms of steel that undergo subsequent hot or cold working. For example, in continuous cast steel, the melt is cooled to facilitate partial solidification and passed through rolls while still very hot. The form may be rectangular or round in cross section and as it continues on the path through the rolls, the cross section is altered. See Figure 5.6.

Wrought forms of steel are available from many commercial sources. In various alloys of the low-carbon steels and in the special high-strength, low-alloy (HSLA) steels we commonly refer to as weathering steels.

Steel in the wrought form, can possess different tempering, carbon contents and strength characteristics that can be tailored to the particular desired attribute or industrial fabrication process. For example, there are certain alloys that have undergone special conditioning to improve deep drawing characteristics. Automotive, appliance and other specialized markets require and benefit from the fine grain, tightly controlled nature of these special steel forms.

Unless specifically ordered, the steels used in art and architecture are less stringent. These are considered commercial quality or commercial steels. The exception are the steels requiring structural characteristics. These are designated SS or SQ for structural steel or structural steel quality. Structural steels are designed for minimum strength behavior, ductility and resilience. They have to be able to accommodate impact loads without fracture. These are the steels that are used to hold up our buildings and bridges. They are used in art and architecture, not for their surface quality but for their mechanical behavior.

FIGURE 5.5   Wrought steel railing.
Source: L. William Zahner.

Some of the available wrought forms of steel are shown in Figure 5.7 and include the following:

Wrought Forms
Foil
Sheet
Plate
Pipe
Tube
Wire
Extrusion (hot drawn)
Structural shapes
Bar

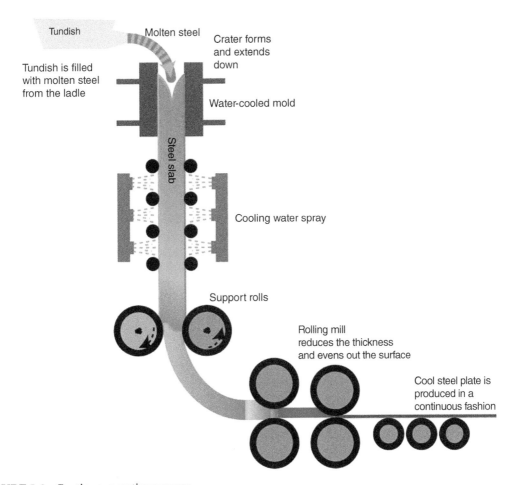

FIGURE 5.6   Continuous casting process.

As identified in the section on alloys, there are two base forms of steel, the wrought forms and the cast forms. There are steel powders and steel shot and bead, but these are used in specialized processes such as cleaning and abrading. They are not normally an end form used in art or architecture.

THE STEEL MILL

All forms of steel begin at the mill source where the steel is initiated. The mill producer takes the charge of recycled scrap and adds in various alloying components, for specific properties and adds other substances to deoxidize the steel. Melting occurs in an electric arc furnace. Once melted, the molten steel is cast into one or several basic forms, blooms, billets, and slabs. The melting and

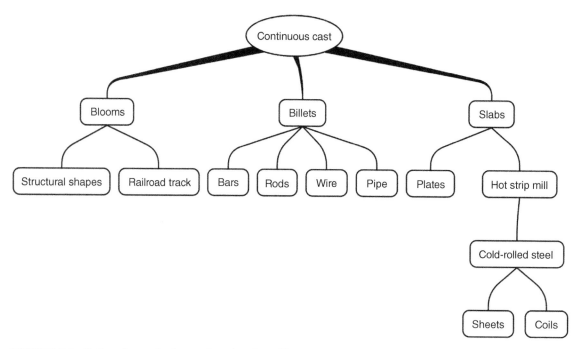

FIGURE 5.7   Paths of wrought forms of steel at the mill.

initial cast is called a heat. A heat will generate a mill certification with a specific heat number for that specific casting of metal. A tracking number is assigned and imprinted on the initial forms. Metallurgic analysis is undertaken on a sample of the heat to verify alloying constituents and physical properties and the corresponding specification, ASTM, ASME, DEN, BS, or other are listed to assure the standard specification has been met. Figure 5.8 shows a mill certification with the information usually provided. Often accompanying the form is a description of the heat treatment used as well as other information associated with various requirements and specifications.

A mill certificate monitors and documents the steel produced in a given heat by a given Mill. The mill certificates follow the steel form as it goes out to the distribution warehouses and to the fabricator or final finishing company of the metal. The mill certification is an important document in the context of quality assurance. The heat may be broken down into different forms, slabs or billets and then further rolled into sheet, plate, wire, or rod but the certification information follows the material and is accessible to the user for verification.

HOT-ROLLED AND COLD-ROLLED SHEET AND PLATE

Sheet and plate, both in the hot-rolled and the cold-rolled surface, are common forms of steel used in art and architecture. It is important to understand the constraints imposed by the normal

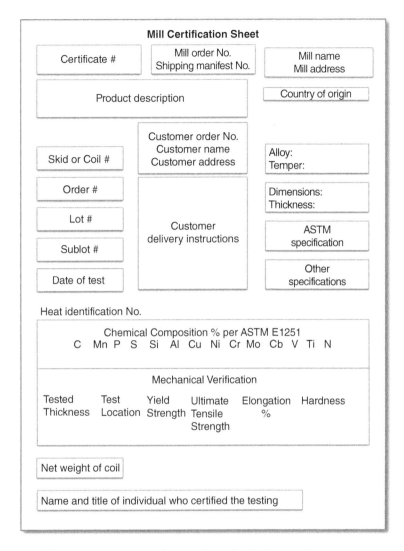

FIGURE 5.8   Mill certificate verifying alloy makeup and mechanical properties.

manufacturing practices. Going outside of these constraints will require design of unique joinery, special processing or accepting less in aesthetic context.

For example, steel sheet like other metals that are passed through reducing and tempering rolls will have different levels of surface quality on the top side (the side facing up) versus the bottom side. This is considered as the prime side because it is the surface more visible for inspection and more apt to be the exposed surface in an assembly. The only time this comes into play when both sides are exposed to view such as perforated screen divider. The nonprime side may have scratches, streaks, and other surface issues that can be an aesthetic concern if visible.

FIGURE 5.9   Typical prime side of hot-rolled steel plate.
Source: L. William Zahner.

The prime side see Figure 5.9 of carbon steel hot-rolled surfaces have visible hues of color usually running the length of the steel sheet or plate. These are common and often, when clear coated for an interior wall surface, desirable.

There are thousands of grades of steels. Specific industries have several specific grades created for unique or minimum design characteristics. For example, grade 250 steel is a carbon steel of nominal yield strength of 250 Mpa. It is used for general fabrication work. It is considered a structural steel.

In art and architecture, the surface quality is more important. The surface quality for steels is not as critical from a design surface and the mills that produce steel plate and sheet as well as the distribution centers do not go to extreme measures to protect the surfaces as they would with copper alloys, stainless steels and aluminum. Steel is more rugged and considered a rugged metal. That being said, the steel is kept dry, often oiled, and protected against damage on wooden skids.

ASTM International, the standard setting organization headquartered in the United States reevaluated the quality designations used in describing the carbon steel sheet and plate materials. Old qualifications have been combined, eliminated or replaced by a newer system that incorporates modern nomenclature and production needs. The quality descriptions were revised in 1996, but many are still being used in referencing steel. See Table 5.1.

These quality codes or designation go further. For example, HSLA...Type B steel can be designated as

HSLA – Type B SCS (**S**mooth, **C**lean **S**urface)

HSLA – Type B PO (**P**ickled and **O**iled)

HSLA – F (Improved **F**ormability)

TABLE 5.1   Old and current designations for steel sheet and plate.

Old designation	New designation
CQ Commercial quality	CS Commercial steel
CQ – low carbon and degassed	CS – Type A
CQ – low and medium carbon	CS – Type B
CQ – high phosphor	CS – Type C
LFQ Lock form quality	FS Forming steel
	FS – Type A
	FS – Type B
DQ Drawing quality	DS Drawing steel
DQ Drawing quality	DDS Deep Drawing Steel
DQSK Drawing quality special killed	EDDS Extra Deep Drawing Steel
SQ Structural quality	SS Structural Steel
HSLA High-strength, low-alloy	HSLA High-strength, low-alloy
	HSLA – Type A
	HSLA – Type B
	SHS Solution hardened steel
	HTS High temperature steel

HSLA steels include the weathering steels. For these, you may want the smooth, clean surface (SCS) designation. These surfaces have the scale removed and are free of oxides and protective oil. They may have some superficial surface rust, but this is not an issue for a steel usually intended on developing a surface rust. If the order is for pickled and oiled (PO), then the sheets or plates will have mill oil that can slow the weathering process.

The HLSA comes in structural grades, such as Grade 50A. It can also be called Grade 350A for the metric equivalent. The 50 stands for minimum 50 ksi steel and the equivalent to this is 350Mpa in metric. This steel's initial appearance is the same as structural A36 steel plate. It is designed for

FIGURE 5.10   Weathering steel grade 50A plates.
Source: L. William Zahner.

higher strength and corrosion resistance and is considered a weathering steel. Figure 5.10 shows HSLA Grade 50A plates in storage.

The steel designations are a code understood by the steel users and warehouses that supply the steel to industry. From an art and architecture context, the carbon steel used would be CS, commercial steel of an available low-carbon alloy. For these steels you may wish to start with the PO, pickled and oiled version. This will provide a better surface finish than say DS, drawing quality.

Cold-rolled carbon steels in the CS designation will have a good surface finish as well. Table 5.2 shows in descending order quality levels on C10080 or C10010.

Appendix E provides several examples of specifications to arrive at good surface quality in sheets and plates of different alloys for use in art and architecture.

TABLE 5.2   Various levels of surface quality.

Commercial bright	Cold rolled in polished and ground rolls
Extra bright matte	Cold rolled in lightly blasted rolls
Cold rolled CS Type A	Cold rolled – matte appearance
Cold rolled pickled and oiled	Cold rolled, pickled in acid, and oiled
DS (Drawing steel)	Cold rolled, surface not critical
Cold rolled CS Type B	Cold rolled, surface not critical

Both the hot- and cold-rolled sheet and plate begin at the same place, the hot slab rolling mill. The slab of steel is reheated and passed through hot rolls to thin and elongate the slab into a thick ribbon of hot-rolled steel. The process rolls the steel hot at a temperature greater than 926 °C (1700 °F).

The surface of the thick, hot-rolled steel plate is pickled to remove scale that forms in this hot-rolling process. The finish on hot-rolled sheet and plate is not as refined as the finish produced after passing through cold pass rolls. The surface of the elongated slab is ground, and the edges are trimmed making it more a thick ribbon of steel. This thick ribbon is coiled into large, rough coils and placed into an annealing chamber to soften the metal. The coil is further hot rolled and passed through a pickling line to remove more scale that forms in the hot-rolling process. The clean, thick, hot-rolled ribbon of metal is recoiled or flattened and cut to length as hot-rolled plate. At this point, it may be oiled to aid in corrosion resistance or left dry. Figure 5.11 shows steel coils awaiting processing.

The hot-rolled and pickled coil can also be transferred to cold rolling stations to be turned into sheet or strip. The temperature of the steel is less than 926 °C. In this case, the coils of thick steel are sent to cold reducing rolls where the metal undergoes further processing. The cold rolling mill passes the ribbon under a set of small-diameter rolls under a tremendous amount of pressure. The metal is rolled as it cools through the reducing rolling mill. The reducing mill is known as a Sendzimir rolling mill, also referred to as a z-mill. This high pressure set of rolls thins the ribbon to the thickness specified. As this happens the grains elongate, and internal stresses develop. The metal is passed through an annealing oven to establish the temper. Modern systems use a continuous annealing line, to adjust the temper and allow further reducing.

At this point the now-thin ribbon is bright and clean of oxides. The surface is smoother than hot-rolled surfaces and the dimensions are set to tighter tolerances. The metal is at a particular temper and possesses defined mechanical properties. The metal can receive a skin pass to produce a

FIGURE 5.11   Steel coils.
Source: L. William Zahner.

smoother surface still with a higher luster or left as fabricated by the rolling operation. The ribbon is recoiled and stored as coil or it can be slit to smaller strip coils or they can be de-coiled and sheared to arrive as sheets skidded to lengths. Here again, the surface may receive a thin layer of mill oil to resist corrosion while stored and handled.

From the mill, the metal is sent to a stocking warehouse or to an end user. In either case they are provided to specification and the mill certification will travel digitally and in paper form as the order travels from mill to warehouse. This will trace the metal from start to finish and is an important control document to affirm that what was ordered was delivered.

A fabricator and manufacturer of steel parts and assemblies for art and architecture should accept only the best surfaces from the mill or more often, from the distribution warehouse. Steel should have scratch and pit free surfaces with the specific mill finish applied to the surface. The surfaces should be free of scale or oxides and have a consistent appearance. They should be well protected, dry and clean.

Mills will produce large plates to specification for various industries. The plates, sometimes referred to as mill plates, are large – several meters in length and as much as 2.5 m in width. The mill will also produce coils of common and popular thicknesses used in industry. These are kept at intermediate processing warehouses along with tubes and bars as inventory for sale. The alloys are determined by market needs and the finishes are normally the mill finish surface produced by the temper rolling.

There are different surface quality standards, as described in Chapter 3. Commercial steel or commercial quality, as it is still often referred to, designated by CS or CQ, will be different for different Mills. It is recommended to acquire representative samples of the surface.

Figure 5.8 shows the information that a mill certification will contain. It has three main sections. The top designates what mill produced the original casting of the metal and contains the original certificate number that travels with all forms of the metal as it moves from the first solidification form such as the billet or slab. The next section corresponds to the entity that ordered it and to what governing specification it was ordered to as well as a description of the form and the finish it was ordered to. This could indicate coils and weights of coils, and it will show the Heat No. that the product was created in. This heat number is important. It is the stamp of a particular melt at a particular time.

The last section of the mill certification indicates the chemical makeup of the material and the mechanical properties. The mechanical properties can relate to a temper used to achieve the particular strength properties. Sometimes for coils they will list the mechanical properties of the lead in portion of the coil and the end portion of the coil. Essentially, they shear off a section at the beginning and the end of the coil and test its strength and measure the elongation. There should be a reference and a stamp, or a signature of which facility and which metallurgist certified the results.

This is an established practice and an important assurance of the metal ordered for the project. Without this, you cannot be certain of what you really have unless you perform the in-depth analysis of the material. If something does not appear right, you can trace this back to the source.

MILL PACKAGING REQUIREMENTS

All steel-fabricated mill products, plate, sheet, coils, rods, and wire are carefully handled and stored to prevent undo corrosion in the form of rust on the surface. Superficial rust on the edges or the surface is common, particularly on nonoiled steels or steel that has been stored out of doors for a short time. The edges usually do not have the protective oil, so they are the first to show the red/orange corrosion products.

Thin steel sheets and coils are stored inside to protect from corrosion. Often, coils are wrapped in what is called *corrosion intercept* wrapping. For coils and stacks of thin sheets, you do not want to get moisture in between the wraps of metal.

Steels are not usually afforded the extra effort of protection that more architectural finish materials such as the stainless steels, aluminums, or copper receive. For the most part, steels are destined for less-aesthetic domains. The skids used to support decoiled plate and sheet are solid, flat, and nonmetallic. Fasteners used to hold the skids together cannot and should not be protruding, and steel banding should have a plastic or cardboard edge protector between the banding and the sheet or plate. Rods and bars are stacked and protected to eliminate scratches from handling.

Storage out of doors should be minimized. On large, thick, structural shapes there is no way around storing out of doors. Corrosion will form on the surfaces, often at scratches and cut edges where moisture can accumulate and access the base metal free of surface oxide. Light corrosion product, uniform oxide is not a concern because usually the surface is destined to be abrasive blasted and coated. However, if the steel has been allowed to remain out of doors for a period of time, particularly in chloride or polluted environments, it can develop more severe corrosion. Corrosion that is deeper and flaking from the surface can be more damaging and replacement should be considered.

Transportation and Storage
- Dry
- Surface and edge protection
- Flat, smooth, strong pallets
- Coil edges protected
- Full wrap of coil and sheet pallet stack
- Use of corrosion intercept paper

SHEET

One of the most commonly used form of steel in art and architecture, particularly when considering a surface, is the sheet form. Sheet forms of steel would be those used as interior cladding and those used as weathering steel exterior cladding. Sheet material is usually coated in a protective oil as it is decoiled, leveled, and cut to length. It can be supplied in unoiled dry form.

Steel sheet is considered thicknesses of 4.76 mm (0.1875 in.) or less. Note that sheet steel is available as both hot rolled and cold rolled. It used to be that plate was considered hot rolled to a minimum thickness of 4.76 mm, and from there the metal was cold rolled to thinner sections. Now, modern mills can hot-roll sheet to a minimum thickness of 2 mm. There will be a difference in the surface appearance of hot- and cold-rolled steel, so it is important to acquire samples of the surface if they are to be clear coated.

It is important to designate a minimum, maximum, or nominal thickness in inches or in millimeters. Gauges are not defined by industry standards, and the descriptive term *gauge* for steel is now relegated to industry jargon. Those in the industry may refer to a 14-gauge thickness, but this is understood to be a nominal 2 mm thickness. Gauge definitions are not a good way of specifying steels. Designate the thickness in millimeter or the inch equivalent. Most mill producers are moving to the metric designations.

It is important to note that there are tolerances in these values, both plus and minus from the nominal thickness. The mill producer wants to maximize the production of material, so they push to the low end of the allowable range that industry has established. That being said, the supply of steel should not fall below the minimum range established by industry.

In North America, ASTM A1008/A1008M-18 establishes standards for allowable thickness tolerances for cold-rolled steel sheet. ASTM A1011/A1011-18a establishes standards for hot-rolled steel sheet. Appendix F is a further list of standards used in the steel industry for hot- and cold-rolled steels.

These standards are aligned with widths of sheet material. The wider the sheet the greater the tolerance allowed. For example, on a 2 mm thick steel sheet of widths up to 1.3 m will have a plus or minus thickness tolerance of 0.1 mm. Over this width and the thickness tolerance is plus or minus 0.11 mm. Whereas a thicker sheet, such as a 5 mm thick sheet, the tolerance on wide sheets greater than 1.3 m in width, will have a tolerance of plus or minus 0.19 mm.

Thicknesses are measured along the longitudinal edge of the steel sheet or plate approximately 75 mm (3 in.) in from the edge.

Some processes of coining and embossing will alter the measurable thickness of the metal. It is the initial thickness before the coining or embossing that defines the steel not the final thickness of the processed metal.

For steel that has been zinc coated by hot-dip galvanizing, or what is referred to as galvannealed[2] is covered under ASTM A653/A653M.

Weathering steel sheet is available in 1220 mm (48 in.) and 1524 mm (60 in.). Thicknesses are the same for the steels, as shown in Table 5.3.

---

[2]Galvannealed refers to a process where the steel is hot-dipped in zinc, similar to galvanizing, but before the zinc has solidified the sheet is passed through a blast of air to remove excess, still-molten zinc. This sheet then is placed into annealing oven where the steel and zinc are allowed to intermix creating a zinc-iron layer. This produces a surface with less zinc but smooth, iron-rich alloy at the surface. This gives galvanized protection but also gives a surface smooth and receptive to paint applications.

TABLE 5.3 Steel sheet thicknesses.

mm	in.	Closest Gauge
4.76	0.187	–
4.55	0.179	7
4.20	0.165	8
3.81	0.150	9
3.43	0.135	10
3.05	0.120	11
2.68	0.105	12
1.91	0.075	14
1.52	0.060	16
1.20	0.047	18
0.91	0.036	20
0.76	0.030	22
0.61	0.024	24
0.46	0.018	26
0.38	0.015	28

COILS

Sheets begin in the coil form, so lengths are cut to order but the standard lengths are dictated by the cutting and forming operations to follow and quality limitations of the end product.

For example, most fabrication facilities have equipment established to cut, form, and process lengths of maximum 3660 mm, or 12 ft. Some are equipped to handle larger sheets, but the average press, waterjet, shear, and laser have limits based on the economics of making the equipment and handling the material itself. Steel should not be dragged across a bed to be scratched or lifted in such a way to induce a kink in the metal. Scratches and gouges in the surface can damage the aesthetics but also provide an anchor for corrosion. A kink is a permanent buckle induced into the metal surface from mishandling. It cannot be removed, because the metal has plastically deformed in this local area.

Limitations of size come into play, not simply for economic reasons, but for practical reasons. In addition, shipping skids for transferring sheets cut to length are aligned to these lengths as well. Otherwise, larger, specialized skids requiring specialized handling equipment will need to be utilized. Simply joining two skids together to make a larger skid can induce distortion into the metal due to uneven or flexible skid supports.

One other parameter that plays an important role in the length of a sheet is *flatness*. Steels hold residual stress from the coiling operation. The process of making the thin ribbon under pressure and coiling induces variable stresses in the ribbon of metal. When removed from the coil, the sheets undergo a stress reduction process, called a leveling line that reduces these stresses by distributing them across the sheet and this flattens the sheet.

You can obtain coils of metal that are several hundred feet in length, but the metal will need some posttreatment to make it useful for architectural or art use. A modern decoiling line will always incorporate a leveling line in the process. These lines are programmable to cut the material to a specific length, remove the internal stresses in the sheet, coat one side with a protective film, and stack the sheets onto skids for storage and shipping.

Coil processing is an efficient method of storage, handling, and transport of steel sheet both at the mill and at secondary processors. The coils can be slit to narrower dimensions and recoiled. Coils can be cut to length or be engaged into a stamping or blanking line, thus reducing waste and improving handling. Often, large master coils are produced at the mill and shipped to be uncoiled and recoiled into smaller coils for easier handling and storage.

The master coil goes to a decoiler that pays out the metal ribbon to a tensioning device that removes the crown in the ribbon of metal. The center of the ribbon on a master has a slight crowning that can be set into the metal and must be tensioned for removal which improves flatness. Leveling lines may be used as well. These differ from tensioning in the way they stretch the metal to correct wavy edges or center distortions. Decoiled ribbons of metal can have an abnormality known as edge wave. Edge wave occurs when the outer extremes of the ribbon are slightly longer than the center of the ribbon. Distortions in the center occur if the center portion is slightly longer than the outer edges.

As the master coil is decoiled, tensioned, and leveled, it travels to a slitting line to be slit down to narrow coils, or it can be decoiled and recoiled into smaller coils or it can be cut to length and skidded into sheets. This is done at the mill or at a secondary plant that handles slitting and decoiling processes.

Modern systems incorporate quality inspection criteria to identify coil breaks and chatter that can be induced into the ribbon of metal as it is processed. Coil breaks should be culled from the line and recycled. For steels this is not always a major issue if they are going to be coated with an opaque paint. If coated with clear coating to show the surface color and tone, these marks can be an aesthetic issue. Culling out the coil breaks and stops should happen at the mill or metal service center during the decoiling but often occurs at the end manufacturer of the product. Here, this quality issue is identified and scrapped. Coil breaks are conditions where there is a mark running across the width of the metal. See Figure 5.12 for this defect.

They appear as lines running the width of the ribbon of metal. Coil breaks develop in a ribbon of metal as it cools. The metal undergoes localized annealing, and as it is uncoiled and then recoiled, this creates stretching in the ribbon of metal. A distortion across the metal is imparted.

Other maladies that can manifest in sheet material are *chatter marks*. These are lines, sometimes very subtle lines, that cross the sheet. They are induced in the coiling and decoiling process where the force is not consistently applied, and the metal is tugged, relaxed, and tugged as it is uncoiled. Adjustments to tensioning of the ribbon of metal may not be consistent from one end of the coil

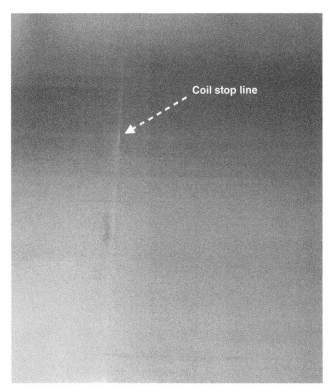

FIGURE 5.12   Coil break mark-in sheet.
Source: L. William Zahner.

to the other as the diameter of the coil decreases. Chatter marks can also occur when finishes are applied to the coil material. Any change in tension can create localized polishing inconsistencies.

Coil breaks and excessive chatter marks should not be acceptable material for art and architecture projects. Quality assurance practices at the mill and then again at the fabrication facility should be able to cull these maladies from ever making it to the finished product. There is no repair process that can overcome this condition if it is present.

The widths of sheets are the other limitation. The widths are limited by the mill's ability to roll and apply pressures to reduce the thickness from the hot form to the thin form. These limitations are physical constraints. Significant pressure is applied to a small-diameter roll called a Sendzimir mill or Z-mill. The Sendzimir mill has 20 larger rolls applying pressure to a set of small, 40 mm diameter rolls in order to squeeze the ribbon to a specific thickness. The wider the sheet or plate, the more difficult it is to apply even pressure and arrive at consistent stress across the metal plate. Widths of steel are limited by production processes and ultimately by the Z-mill.

All subsequent handling and processing operations are designed around these limitations. Coiling and decoiling, flattening, packaging, and laser-cutting beds are all designed to accommodate most of these widths. Custom-slitting operations can take the standard widths of a coil and slit them down to requirements of the project or fabrication facility. Pricing is higher because the off fall is recycled as scrap. You will be charged for the original coil width, so the unit price on off-size coils will be higher.

Carbon Steel Typical Widths

2438 mm	96 in.
2000 mm	78 in.
1524 mm	60 in.
1500 mm	59 in.
1219 mm	48 in.
1000 mm	39 in.
914 mm	36 in.

On coils, only the directional satin finishes, embossed finishes, and coined finishes are currently available. Application of other finishes will become available as process improvements overcome the limitations of applying these finishes on a continuous basis.

PERFORATED AND EXPANDED STEEL

Steel sheet can be perforated and made into expanded metal. A necessary condition to overcome, however, is the potential of corrosion on the carbon steels. The weathering steels with their superior corrosion resistance are less of a concern and will perform well out of doors. In particular, when the perforated or expanded metal acts as an exterior screen, the front and reverse side will show similar development of the oxide.

PERFORATED STEEL

Perforated steel is used as screening material and as decorative panels. All sheet forms and finishes can be perforated and custom perforated. There are some limitations to width of perforated steel based on the equipment used to create the holes. The overall width is dependent on the machine width that produces the perforation by punching out holes and shapes or by laser, plasma, or waterjet tables for custom perforating. Length is less an issue; however, there exists some practical limitations of handling and subsequent forming operations.

It is important to note that in sheet metals, one side is prime or considered the face side, while the back side of the sheet does not receive the same level of care. On perforated surfaces, the back side will be less ideal and may require some additional finishing if it is to be exposed to view. Often, the process of perforating can scratch the reverse side as the sheet moves across the bed or is passed through a set of piercing rolls. You can protect the reverse side by a PVC layer, but this will add cost both in the protective film and in the time needed to remove the film.

Sharp tools and higher power are critical for piercing steels with punching machines such as turret punches or other CNC-controlled punch presses. Steels, particularly the HSLA steels that make up the weathering steels are harder and tougher than other metals. Dull tools induce stress into the steel as they tend to rip the edge rather than a clean shear. Treated and coated punching tools help reduce heat during the piercing operation.

When perforating steel, there are several important relationships to keep in mind. For round holes, the ratio of the diameter to thickness is critical. A ratio of 2 to 1 for steel will reduce inducing stress into the sheet and reduce stress on the tool used to punch. For other shapes, this ratio would correspond to the smallest of the dimensions. For example, if the shape is rectangle, the ratio would be 2 to 1 for the shortest distance to the thickness of the sheet. For example, if the sheet thickness is 2 mm, then the shortest distance on the rectangle should be no less than 4 mm.

The clearance and die design are critical as well. The tool and die must be designed for the correct clearance of the steel thickness. The die should be designed with rounded corners instead of sharp corners when piercing steel. Sharp corners will damage the die and can create small rips in the corner of the piecing. Figure 5.13 shows a custom-perforated, punched surface made from 3 mm thick preweathered steel.

Perforated metal is extensively used in architecture and art. Perforations using punching operations often are symmetrical. Circles, squares, rectangles, even stars and shell shapes. It is not an absolute requirement that the shape is symmetric. Some of the nonsymmetrical shapes can get caught in the tooling. Cutting perforations with a laser affords an infinite variety of shapes. Here again, the nonsymmetrical forms and even symmetrical forms smaller than the slats used to hold the steel can twist and interfere, slowing the process down.

Perforated steel sheet is available in the same dimensions as plain carbon sheet. Prepainted and pre-hot-dipped galvanized sheet can be perforated but the edges of the holes will be bare carbon steel.

FIGURE 5.13 Washington Elementary school custom perforated steel. Designed by HMC Architects.
Source: Courtesy of ImageWall.

FIGURE 5.14   Custom waterjet cut plate.
Source: L. William Zahner.

Post painting perforated carbon steel and post galvanizing of perforated carbon steel is possible but not always practical. Small hole sizes, less than 9 mm (0.375 in.), are subject to encroachment on the holes, filling the openings with irregular paint and zinc.

Perforated metal is normally produced on sheet or thin plate material. Custom perforation can be accomplished with waterjet on large, thick plates. Figure 5.14 shows a custom-perforated 12 mm thick plate made from weathering steel.

EXPANDED METAL

Steel is well suited for expanded metal. Expanded carbon steel or weathering steel sheet offers strength along with significant economy. Figure 5.15 shows shelving made from expanded carbon steel sheet, then powder coated. The shelving was load tested to prove out the strength of the expanded sheet metal. Expanded metal involves slitting the steel at preestablished intervals and then either upsetting the area at the slit to cause one side to extend out and the other side to extend inward or the metal is pulled and stretched to open the slit into a diamond-like perforation. No metal is removed in the process and by expanding the material, better coverage can be obtained. This allows for a better yield of metal coverage. As the metal is slit and stretched the overall area increases drastically.

There are various sizes and configurations of expanded metal. Typically, the opening made in the expanded metal have a slightly raised "lip" created as the stretching occurs. This form is known

FIGURE 5.15 Drop box expanded metal.
Source: Courtesy of United Factory Productions™.

as *standard expanded metal* or *formed expanded metal*. This formed sheet can be further processed to flatten out the lip, referred to as *flattened expanded metal*.

Expanded steel sheet can be ordered with a diamond pattern running lengthwise or across the sheet. The diamond pattern can be staggered or straight line. The diamond can be created in a multitude of sizes, from very small on thin sheet to very large.

Other decorative patterns are also available from different manufacturers of expanded metal. Availability, sizes, and thicknesses of more custom openings should be investigated directly with the manufacturer.

Expanded metal is usually ordered by the "Long Way of the Diamond," LWD, and the "Short Way of the Diamond" SWD, as shown in Figure 5.16. The width of the strand, or metal connector, is

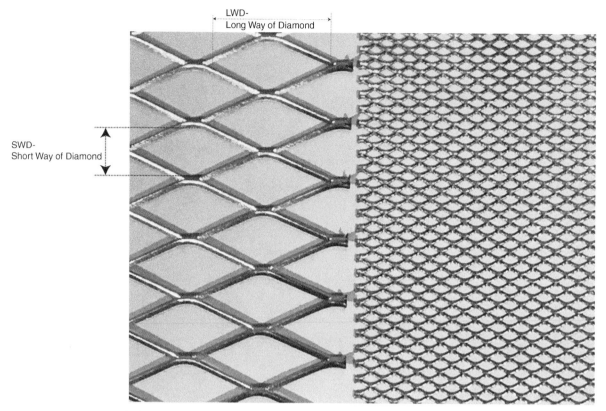

FIGURE 5.16   Expanded metal design criteria.
Source: L. William Zahner.

also qualified, along with the metal thickness. Expanded metal comes in many sizes and dimensions. Even very small-mesh sizes are possible with thin steel to create screens coverings.

Expanded metal can be painted or hot-dip galvanized for corrosion protection. It can also be made from HSLA weathering steel. Normally, expanded metal is produced from cold-rolled steel, so the thickness is limited to 3 mm or less.

PLATE THICKNESSES AND WIDTHS

Steel thicker than 4.76 mm (0.187 in.) is considered plate. The main distinction between steel plate and steel sheet is the finish and refinement of the surface. Sheet steel is passed through the Z-mill to reduce the thickness, and subsequent passes are through cold rolls. This elongates the grains and smooths the surface. Figure 5.17 shows various plates of steel and how they are normally stored. Because steel is considered more an industrial material than an architectural surfacing material, special care and storage are not paramount. This lends to the economy of the material as well.

The plate form of steel may see the planetary rolls, which reduce the thickness. Plate is rolled hot, so the finish surface is not refined to the level of cold-rolled steels. The plate can have inclusions

FIGURE 5.17 Steel plates.
Source: L. William Zahner.

on the surface and waviness across the surface. Thin plate can receive cold passes similar to sheet and the surface quality improves but still distinguishable by a coarse surface. As the thickness of the plate increases, the surface of the plate is not flat and there is often pits and inclusions apparent on the surface (Table 5.4).

Plate widths are similar to sheet. Depending on the source, plates of various thicknesses are available in widths up to 3 m (120 in.). Standard thickness is shown in Table 5.5. These are nominal with European and Asian suppliers rounded to the nearest millimeter.

TABLE 5.4   Finishes on plates.

Finish form	Description
Hot rolled	Scale on the surface; very rough
Hot rolled and pickled	Scale removed; rough surface
Hot rolled, pickled, and temper passed	More refined surface; smoother
Hot rolled, pickled, and cold rolled	Improved surface with fewer inclusions

TABLE 5.5   Standard widths of steel plates.

mm	914	1000	1219	1524	1829	2438	3048
in.	36	39	48	60	72	96	120

Plate lengths are limited by handling and transport. Plates cannot be leveled in ways sheets can. They can be rolled to even out the stress in thinner sections if the rolling facility can handle the weight of large steel plates. Plates are not as flat as sheet forms of steel and the surface will have variations due to cooling from the hot form. Plates passed through a cold roll have a smoother, more refined surface. In all cases, though, architectural surface finishes produced for plate forms of steel are not to the finish level obtainable in sheet forms of steel (Table 5.6).

WEATHERING STEEL PLATE

Weathering steel is available in the following widths:

1220 mm	48 in.
1524 mm	60 in.
1830 mm	72 in.
2440 mm	96 in.

Other widths are available in special order. Lengths are determined by handling and processing equipment and space. Plates are limited more by final fabrication processes than by mill manufacturing limits.

Thicknesses of weathering steel plates are from 4.75 mm (0.187 in.) to as thick as 127 mm (5 in.). The thicker the plate, the rougher the surface finish.

TABLE 5.6   Plate thicknesses.

mm	in.
4.8	0.188
6.4	0.250
8.0	0.313
9.5	0.375
11.1	0.438
12.7	0.500
14.3	0.563
15.9	0.625
22.2	0.875
25.4	1.000
31.8	1.125
38.1	1.500
44.5	1.750
50.8	2.000
63.5	2.500
76.2	3.000
82.6	3.250
88.9	3.500
95.3	3.750
101.6	4.000

BAR AND ROD

Bar and rod forms are solid, homogeneous steel alloys and are produced in stock lengths by hot rolling, cold reducing, forging, or extruding. Maximum lengths are what can be handled and transported. For bar material the thickness is 3 mm (0.125 in.) or greater and widths to 254 mm (10 in.). Widths in excess of this are considered miniplates. Rod material is available in rounds (circular cross section), octagons, and hexagons. Minimum dimensions in diameter are 3 mm (0.125 in.).

The mill finish is a dull, pickled, scale-free surface. For art and architectural uses, the bar material often goes through cold-finishing processes.

Bar material goes through a process known as turning to remove defects on the hot-rolled surface prior to further finish operations. Turning is a milling operation that cuts away the surface until a specific dimension is arrived at. Another process used is grinding, also called centerless grinding, where the bar material surface is ground down to a certain dimension. This arrives at a surface better suited for further finishing.

There are many variations on standard bar and rod that are produced by cold drawing through a die. These are limited in dimension and require significant quantity to account for the die cost, but they reduce machining time and provide an excellent surface.

Other custom shapes can be hot rolled. The finish is tougher and requires subsequent pickling to remove scale. Hot rolling is used to produce long, straight forms that will undergo subsequent welding, shaping, or forging operations.

The edges of hot-rolled sections and bars are slightly rounded. The dimensional tolerances are less. Cold drawing bar and rod increases the yield achieved by the mill and reduces surface imperfections that are often present in hot-rolled sections. Hot-rolled sections also shrink as they cool, which requires greater dimensional tolerances. Hot-rolled is more economical than cold-rolled bar (Figure 5.18).

Cold-drawn steel is a precision reduction process used to develop a linear shape with a precise cross section. A bar or rod is passed through a series of dies that shape the metal into a precise cross section. The benefits are a very accurate cross section with very good surface qualities. The process produces near net shapes with little waste.

The cross section is the limiting factor as well as the quantity minimums. The shape will need to be able to fit within a circle of diameter less than 75 mm (3 in.) and not be overly complex. The process needs a minimum quantity dependent on the cross section configuration.

Rectangular is available in several common alloys. C10180 and C10450 are standard, as well as A36 steel alloys.

A36 alloy rectangular bar is available in several standard sizes. The following is the range of thicknesses by widths.

Bar thickness range   **Bar width range**
3–60 mm   12–to 305 mm

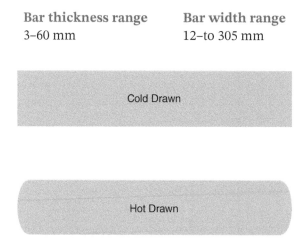

FIGURE 5.18   Edges of cold and hot drawn bar.

TABLE 5.7 Nominal diameter of round rod.

mm	in.
4.78	0.19
6.35	0.25
7.95	0.31
9.53	0.38
9.98	0.39
12.7	0.50
15.00	0.59
15.88	0.63
19.05	0.75
22.23	0.88
25.40	1.00
31.75	1.25
34.93	1.38
38.10	1.50
44.45	1.75

Square bar is available in A36 alloy from 9.5 mm square to 150 mm square (Table 5.7).

ROUND ROD

Common available alloys:

C10180
C10450

TUBING AND PIPE

Steel piping is produced in significant quantities for the oil, gas, and chemical processing industry. Pipe is distinguished from tube in that pipes are vessels for transporting fluids and the inside dimension is critical for what is being transported. The outside diameter and pipe thickness are nominal values. Pipe is provided from the mill with a hot-rolled and pickled surface. No particular care is extended to this normally industrial surface beyond removing scale and providing a clean surface.

FIGURE 5.19   Sculptural light form created from round steel tube.
Source: Artsists Julia Cole and Leigh Rosser

Tubes, on the other hand, are structural, and the outside dimension and wall thickness are exact values. Tubes are cold worked to achieve accurate dimensioning. They can be seamless, produced by hot-rolling processes with initial surfaces similar to a hot-rolled surfaces requiring further finishing. Tubes can also be created from hot-rolled plate, cold-rolled sheet, and strip shaped in rollers to the tube form and welded along the length. The weld is ground down and prepared to reduce appearance effects. This occurs on both the inside and outside surface.

Tubes can be drawn on a mandrel to very tight tolerances. Figure 5.19 shows an art form created from steel tube.

Pipe is always round in cross section, while tubing can be round or rectangular in cross section. Tubes are commonly available in the following alloys:

C10100
C10150

C10180

C10260

C10450

Tubes are available in diameters of 12.7 mm (0.5 in.) up to 308 mm (12 in.).

Wall thicknesses are from 1.6 mm (0.65 in.) to 12.7 mm (0.5 in.). Subject to availability.

STRUCTURAL SHAPES

Steel structural shapes are common in the building industry. Steel structural shapes are created by hot forming, cold forming, extrusion, and welding. The hot-forming process is performed at the mill. It is an amazing process of rapidly forming the red-hot steel block or rod form into a tube, I-beam, or channel shape. While hot, the form is created by a series of dies, each die adding incrementally to the shape.

The shapes produced this way have rounded corners and edges. Structural steel is ubiquitous when we think of steel. Figure 5.20 shows examples of structural steel used in building construction.

Welded sections are constructed from hot-rolled plates. These are created by welding the flanges to create the shape. They provide economy and are similar to the bolted plate girder. These are created to a specific mechanical requirement. They are distinguished from the hot-rolled shapes by the lack of rounding at the corners and edges. Often, they are welded on one side of the intersection of the plates.

The finish provided on steel structural shapes is the hot-rolled surface. The surface can be prepared for painting by blasting the surface with abrasives. Structural sections can also be galvanized.

Steel structural shapes can be produced by laser fusion of plates to form the shapes. The laser-fused shapes are made from fused plate sections. The finish on the laser-fused sections is superior and can be finished readily for art and architectural use. The corners are square as the plates are structurally joined.

The structural steels fall under ASTM A36 minimum requirements. The HSLA steels, also called weathering steels, are available in the various structural shapes. These are commonly used in structures where they are exposed without paint coatings and maintenance is minimal. Many bridge structures are made from these HSLA steels.

HSS: HOLLOW STRUCTURAL SECTIONS

Round tubes, rectangular tubes, and square tubes, referred to as HSS or hollow structural sections, are common in the building industry. They are also referred to as CHS, RHS, and SHS for circular, rectangular, and square hollow sections, respectively. These are hollow sections made from structural steel designed to meet minimum strength requirements, as designated by ASTM A36/A36M. These shapes are hot formed from plate with wall thicknesses of 12.7 mm to as much as 50 mm.

FIGURE 5.20   Structural shapes.
Source: L. William Zahner.

The thickness and outside dimensions of the hollow member are determined by the structural requirements. The corners are always rounded. There is an apparent welded seam running the length.

Hollow structural forms are common in the building construction industry, particularly where loads are applied from different directions. On occasion, they are filled with concrete to add further strength to a column form. This is known as a *lally column*.

In architecture, these shapes may be used to support wall assemblies or other forms, but in most instances, they are not considered aesthetic features and are concealed. In art, these can be used as supporting members, or conduits for concealing electrical wiring while still providing strength to the form.

The surface is not particularly pristine. The weld running the length is visible and not ground smooth. The corners can show tooling marks running the length of the members. They can be galvanized to improve corrosion resistance. Hot dipping the forms in molten zinc will coat both the inside and outside surfaces. They can be abrasive blasted and painted similar to other structural steel members.

WIRE, WOVEN WIRE, AND SCREENS

Carbon steel wire is not typically used in art and architecture. It is not as durable or aesthetic as its cousin, stainless steel wire. Still, it does have useful attributes of cost and strength. There are a number of carbon alloys, low-, medium- and high-carbon content, used to make wire. Different tempers and strengths are available for the specific use.

Carbon steel barbed wire created in the late 1800s and throughout the 1900s was a mainstay in the movement west in the United States. First patented in 1867, it was extensively used as fencing material and was capable of restraining cattle. Figure 5.21 is an amazing sculpture on the Montana State campus, made from barbed wire and steel supports. The artist, Jennifer Pulchinski, has made an art form of recycled barbed wire found across Montana.

Steel is available in multiple wire forms, both single strand and multiple strand. Single wire is available in diameters 0.254 mm (0.01 in.) up to 12.7 mm (0.5 in.). Multi-strand cable can be developed from wire into massive assemblies. Steel wire is provided in a semi-finished state free of scale. Wire is produced from rod material and cold drawn to specific dimensions and structural characteristics and tempers from spring temper to fully annealed. Figure 5.22 shows coils of steel wire.

Wire screens made from weaving metal threads on special looms similar to woven cloth are available in a multitude of forms. Simple screening is available in several sizes, called *sieve size*. These can be woven to very tight tolerances. Woven steel is very strong and durable and is used as inexpensive sieves and barriers. It can be easily welded and formed. Similar to perforated steel, tightly woven wire is not easily painted or galvanized without filling the open spaces. More open screens, minimum 9 mm spacing, can be painted and galvanized. Chain-link used as fences is a common galvanized steel wire form.

Woven wire screens are made from round wires, semi-round wire, or flattened wire. The wire can have diameters as small as 0.02 mm. Various sizes and shapes can be intermixed.

FIGURE 5.21 "A Nesting Place," sculpture made out of barbed wire.
Source: Jennifer Puchinski artist.

Steel cable and wire come in various sizes and rated load capacities. The cables are made from wire of various carbon content, depending on strength and stiffness. There are numerous alloys to choose from, but normally the choice of cable is based on mechanical characteristics that have been tested and verified. Steel wire comes in several diameters and load capacity ratings. Fittings can be swedged on to the ends for special attachments and decorative appearances.

Steel cable is subject to corrosion. Moisture can get trapped in the mix of small wires. Because of this, the steel cabling is often coated to keep moisture out or is made of larger, galvanized steel wire.

EXTRUSION

Steel can be extruded. Steel extrusion shapes are used on window and door sections as well as other forms with consistent dimensions and tight tolerances. Sharp edges are eliminated and

FIGURE 5.22   Steel wire.
Source: L. William Zahner.

rounding of the corners and fillets is required. The extruded steel is hot formed with significant pressure. Small structural shapes used in different industries where consistency and strength are paramount.

There are a number of carbon steel alloys extruded. C10080 and C10100 are typical, but other low- and medium-carbon alloys are also available.

The shapes possible are limited in outside dimension; they need to fit within a certain diameter circle, typically no more than 165 mm (6.5 in.). The process of extruding metals goes back centuries.

In the hot extrusion process, the billet of steel is heated to around 1200 °C, slightly higher than the temperatures used on hot rolling. The hot billet of steel is rolled in glass powder that melts and adheres to the outside surface. A glass pad is placed toward the back face of the extrusion die to insulate the die from the hot steel.

Extruding of seamless steel tubes was a major market in the past, but continuous welded tubes are more economical to produce and have for much of the industry, replaced extrusion.

STEEL WINDOWS

Steel window sections are made from custom-extruded steel shapes, steel bars, roll-formed steel sheet, and break-formed steel sheet. When the shapes are formed from steel bar, alloy C10180 is commonly used. Steel windows were the fashion in the early part of the twentieth century until extruded aluminum came into use.

Hot-rolled window sections were originated by the British firm Crittall Manufacturing Company in 1912. Their designs are still in use today. They created a hot-rolled shape for window sections. The shape came in three sizes, depending on the strength needed. Figure 5.23 shows the general shape.

Steel windows provided strength and durability, but corrosion has always been a challenge. Steel window seals would decay and allow moisture into the steel form, where it would slowly corrode. The corrosion would expand and open the joint. The other concern today in restoration efforts is that many of the steel windows of the past were coated in lead paint. See Figure 5.24.

FIGURE 5.23   Geometry of Crittall steel window section.

FIGURE 5.24   Rust expanding on a steel window section.
Source: L. William Zahner.

Steel window sections are making a resurgence due to better paint finishes and the desire for thin profiles. The strength afforded by the steel allows for smaller sections than comparable aluminum. The profiles were modified in the 1960s and later in the 1990s to accommodate energy-saving glass and thermal isolation components. There are several standard cross-sectional shapes made by hot forming and extruding steel components for windows. These are referred to as W20 and MW profiles. Newer, modern profiles incorporating energy-saving aspects of modern multiglazed windows and thermal brakes can be made from roll-forming steel sheet into sections. The economy of roll forming as well as the thermal improvements by adding built-in gasketing, coupled with the recyclability of steel, have brought the steel window into consideration for modern construction. Figure 5.25 is an example of modern steel window framing.

FIGURE 5.25   Modern steel window frames.
Source: L. William Zahner.

POWDER, BEAD, AND SHOT

Steel is available in powder form, small beads, and steel shot, not common in art or architecture. These are used frequently as abrasive blast media and as shot for hunting and targeting. These minute forms are available in low-carbon alloys of no particular grade.

FOIL

Steel foils are used in specialized industries. Carbon steel can be rolled very thin to foil thicknesses and provided with various tempers. The available material is narrow strip and has a reflective surface finish due to the numerous passes through reducing rolls.

Typical foil alloys are C10100 and C10080. Steel work hardens as it is cold reduced into foil thicknesses. The process of producing sheet dimensions to thin foils is difficult and requires annealing steps. The edges of these foils are razor thin, making handling and working with thin foils a delicate process. Foils are used as barriers, reflectors, and a number of other situations that require a durable low-reflective material. Available foil thicknesses are from 0.05 mm (0.002 in.) to 0.5 mm (0.02 in.). Foils are often applied to other backing materials to make them easier to handle.

CAST STEEL

Casting is an age-old process that has changed little in over 2000 years. The biggest difficulty with casting steel is the energy required. To cast with iron and steels, the temperatures are on the order of 1480 °C (2700 °F).

Cast iron and steels can develop intricate detail; however, the finishing and protection of the detail can be problematic. Getting back into recesses to remove slag and oxide and then protecting these hidden regions from corrosion can be difficult to accomplish. Figure 5.26 shows some cast iron "gargoyles."

Cast iron is brittle and cannot be hammered or shaped. It can be machined, though, and often for intricate parts and fittings, the casting will undergo machining to net size dimensions. It is important to not use cast iron forms in structures subject to bending or tensile stresses because they can be prone to cracking.

Castings are produced in sand molds, which can result in a coarser surface. Once the cast steel part is removed from the mold, the surface scale is taken down with clean sand or steel shot blasting. Abrasive blasting is the first step in preparation of a cast steel part. This step removes sand from the mold and heavy scale. After abrasive blasting, the surface is pickled to remove remaining oxide, followed by rinsing and final finishing.

Once the steel casting has been cleaned, it can receive paint or it can be hot-dipped galvanized. All cast forms of iron and steel can be galvanized.

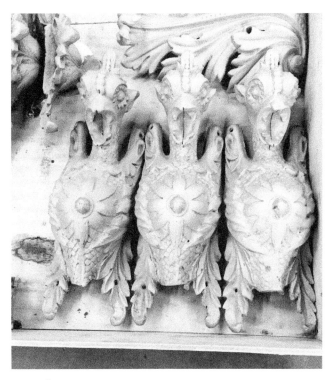

FIGURE 5.26   Cast iron gargoyles.
Source: By Shutterstock.

The selection of the cast process for steel begins with evaluating these three criteria:

1. Finish surface quality
2. Dimensional tolerances
3. Cost

Each of these plays a significant role in determining the value of using cast steel as an option. Finish quality is a function first of the mold used to cast the part and second of the skill of the finishing operation. For cast steel and iron surfaces, the finish is usually rough due to the texture of the sand used. The beauty of casting with steel and iron is the economy, and since most uses of cast iron and steel are industrial, the method most often used is sand casting. Better surfaces can be obtained in other cast methods, such as ceramic shell or investment casting methods that have more refined molds but at a significant premium in cost. Figure 5.27 shows some decorative "faux nuts" and the coarseness of the sand cast surface obtained.

Dimensional tolerance is tied directly to the casting process itself. Small investment castings can have very tight tolerances, as low as 0.1 mm, while large sand casting parts have tolerances of

FIGURE 5.27　Cast iron "nuts."
Source: L. William Zahner.

as much as 12 mm. Machining can aid in achieving tight tolerances, but this will add cost. Cost always plays into design and material choice. In the case of casting, cost is a constraint dependent on quantity, mold process, alloy of steel, annealing rigor and skill level of the operation doing the casting. Modern processes of producing molds from rapid prototyping equipment, coupled with development of ceramic slurry compounds, are significantly improving time and cost.

These are the forms available to the fabricator of art and architecture. They are not unlike the forms other metals are produced in. Their size and shape limitations are defined for the most part by industry production processes at the mill, handling and transportation constraints, and by the equipment used to further fabricate the metal into the shapes and forms of art and architecture.

CHAPTER 6

Working with Steels

Once you learn to look at architecture not merely as an art more or less well or more or less badly done but as a social manifestation, the critical eye becomes clairvoyant.

Louis Sullivan

INTRODUCTION

Steel is an ideal metal for a fabricator of metal products. Much of the fabrication equipment is centered on working with steel, as this amazing metal is woven throughout various industries. It is not the easiest metal to work with, though. Its toughness dulls down tools and requires power to form. Its dust and fine particles can contaminate the surfaces of other metals. Its cousin, stainless steel, wants nothing to do with it. Steel will quickly contaminate the surface of stainless steels if allowed to remain and corrode.

Steel has to be kept dry and clean as it is worked. Then after it has been worked, it must be protected from the elements with coatings unless it is one of the weathering steel alloys. Weathering steel is an enigma when it comes to steels. Weathering steel wants to interact with the environment. Moisture and the heat from the sun enable it to develop a protective layer of rust, unlike other steels.

Steels can be welded using all the welding processes, many of which were developed for steel and adapted to other metals. Welding of steel is one of the most versatile and creative processes available to the steel fabricator. Joining of steel parts by fusing them together in the processes of welding is possibly one of the most desirable qualities of steel to the fabricator. Figure 6.1 shows steel fabrications that are assembled by welding to become monolithic, continuous forms.

Once properly welded and assembled, these fabricated forms offer a consistency in strength and durability. They can be coated with paint to protect the steel from corrosion and provide an

FIGURE 6.1   Top image shows welded staircase being ground to smooth out the welds. Lower image is an entry desk made of steel plate that has been welded and ground.
Source: L. William Zahner.

aesthetic appearance of beauty. The stairs shown in Figure 6.2 were designed by Diller Scofidio + Renfro. Made from formed steel plates, welded and ground, then finish painted, the stair form is nearly 15 m in length.

This chapter covers some of the processes and challenges with working with the various steels for art and architectural projects. Steels are economical and offer versatility. All the steels can be worked in similar fashion. In art and architecture, we will concentrate the discussion on the low-carbon steels and the high-strength, low-alloy (HSLA) weathering steels.

FIGURE 6.2 Julliard stairs made from thick steel plates welded and ground to produce a monolithic form. *Source:* Designed by Diller Scofidio + Renfro.

CUTTING STEELS

There are a number of methods in common use to cut steels. See Table 6.1. For the wrought forms these are computer numeric controlled(CNC) on tables and produce an *x-y* orientation or they are produced by hand or robot assistance to cut the more complex shapes.

TABLE 6.1   Various cutting mechanisms used on steels.

Cutting operation	Complexity	Limitations
Shearing	Simplest process	Straight lines. Sheet and plate
Saw cutting	Straight edges and curves	Rough edge. All forms
Laser cutting	Fast, CNC shapes	Sheet, thin plate, tubing (five-axis)
Plasma cutting	Fast, CNC shapes, manual	Rougher edge, sheet, plate, in situ
Torch cutting	Fast. Hand or machine assist	Rough edge. Heavy plate or forms
Waterjet cutting	Slow, CNC shapes	Frosted edge, sheet, thick plate
Punching	Fast, holes, CNC shapes	Sheet

SHEARING AND BLANKING

Shearing and blanking of steels involves slightly more power in the cutting shear because of the higher shear strength and hardness of steels. The HSLA steels in particular are much stronger and harder than other metals used in art and architecture. Additionally, shearing steel demands addressing the shear blade more frequently, keeping it aligned and sharpened. It wears out faster due to the hardness of the metal.

Shearing is an operation involving two sharpened and hardened steel blades. One blade is fixed, and the other is brought down with such force that a metal strip undergoes severe plastic deformation to the point that it fractures at the surface line where it contacts the shear blades. This fracture propagates through the metal. When there is proper clearance between the cutting blades, the crack that propagates penetrates only a portion of the thickness. One crack forms from the top blade and another crack forms from the bottom blade. These cracks meet near the middle to provide a clean fracture line.

Signs of wear are rough edges where the steel has been cut or there is a slight roll-over of the sheared edge. This edge can be sharp and jagged due to ripping from the shearing blades being out of alignment. Clearance is critical when working with steels. Blanking, punching, and shearing operations on steel can require an increase in clearance of as much as 10% of the thickness of the metal.

Shearing and blanking are typically performed on thinner sheet steel. There are mechanical shearing devices used to shear bar and angle of thin gage, less than 5 mm. These are quick-cutting shears of considerable tonnage.

SAW CUTTING

Wrought and cast steel can be cut with cold saw blades. Steel is readily cut with special saw blades and band saw blades. A cold saw is a steel circular or band type saw. The tips of the blades can be coated in tungsten carbide, coated with synthetic diamond or with titanium nitride. Saw cutting can induce heat into the steel, causing localized distortions along the edge. Usually a coolant spray is applied at the point of the cut to reduce the heat and extend the life of the blade. It is good practice to conceal a saw cut edge in a cover plate or folded under to create a lap joint. It is very difficult to achieve an acceptable saw cut edge in thin steel sheet. Bar, tubing, and piping can be saw cut with acceptable edges along the cut due to the greater section and thicker material.

LASER CUTTING

Steel parts are effectively cut with laser. Lasers are excellent and efficient cutting tools for cutting all steel alloys. Intricate shapes and detail can be cut into steel sheets and plates. A laser works by concentrating a highly focused beam of light onto the metal surface. Electromagnetic energy of the laser light is transformed into thermal energy within the thickness of the metal. The light is absorbed by the metal, and this energizes the electrons. The electrons are accelerated by an energy field within the crystal lattice of the metal, and this generates heat. At the point where the light strikes, the heat is so high that the material melts or vaporizes. The amount of heat generated is based on the light absorption of the material being cut. Once it pierces the metal, the cutting action begins. The beam of light moves along the contour, melting the metal, while a jet of gas blows the melted metal downward, leaving a kerf cut not much wider than the beam itself.

The surface properties of the metal influence the optical behavior of the laser beam. The peaks and valleys of a diffuse surface can trap some of the light and speed up the process. Steel sheet and plate can be cut with gas lasers (CO_2) and with fiber lasers. Fiber lasers are solid state, Nd:YAG (neodymium doped, yttrium-aluminum-garnet) lasers that deliver the energy by fiber. This is a very efficient method of cutting steel.

Higher-power lasers are needed to cut thicker metal. A 2000 W laser can cut 4 mm thick steel adequately. A 4000 W laser can cut steel as thick as 15 mm.

High-powered lasers can cut up to 25 mm (1 in.) thick steel plates without leaving a burr or edge. There are five-axis lasers that can cut steel tubing and pipe as well as forged shapes. These are specialized devices set on robotic arms.

The edge produced on a laser should be clean and free of heat discoloration and burrs from the cutting process. Heat tint is usually not an issue with laser cutting when using a nitrogen atmosphere. If the edges do show discoloration, then it must be removed using pickling treatments. Figure 6.3 shows an image cut into 4 mm thick weathering steel plate by a 4 kW laser. The holes are pierced and cut by the laser; note the variable shapes, particularly along the shaped edge.

Lasers are CNC-controlled cutting devices. They are very efficient cutting tools and can be programmed to cut intricate shapes in both two-dimensional surfaces and three-dimensional parts. If there is a lot of piercings in a given design, some shaping may occur in the surface as heat is absorbed by the metal.

Additionally, there can be discoloration along the area where a laser drill or piercing occurs. Waxing the sheet or plastic coating the sheet can resist some of the discoloration that occurs around the cut. Figure 6.4 is a design cut from a laser and folded, made from custom blackened steel.

FIGURE 6.3   Eagle cut in 4 mm thickness weathering steel.
Source: Courtesy of ImageWall™.

PLASMA CUTTING

Carbon steels are effectively cut with plasma cutting systems. Plasma cutting is typically performed on two-dimensional, CNC controlled tables. The steel plate or sheet is set onto a steel lattice and the high-energy plasma beam cuts through the metal. Plasma cutting involves creating an electrical charged ionized gas and forcing the gas through a small orifice. An electrical current is generated from a remote power source that creates an arc from the steel work piece, which is given a positive charge, and the plasma gas which is given a negative charge.

For steels the gas used is nitrogen or an argon hydrogen mix. Carbon dioxide and oxygen gases can be used to create the plasma jet, but these will create a darkened cut.

Temperatures in the high-velocity plasma jet can be as high as 22 200 °C as this highly charged gas melts the metal and blows it away. There is not a lot of heat transferred to the steel, but if there are a number of piercings you can expect warping of the thin steel. The high-definition plasma cutting systems reduce the heat-affected zone and produce a fine cut line in thin steel. Figure 6.5 is an example of a plasma cut artistic design in steel.

The handheld and less sophisticated systems will have a larger kerf and more oxidation on the edge. The kerf of the cut, which is the term given to describe the edge or cut, is rougher than water

FIGURE 6.4   Pattern cut into blackened steel with laser. Design by the artist Joe Kamm.
Source: Courtesy of ImageWall.

FIGURE 6.5   Plasma-cut design by the artist Jesse Small.
Source: design by the artist Jesse Small.

or laser. There can be a redeposit of molten metal along the kerf. This redeposit is rough oxide that will need to be removed. For art and architectural projects, plasma cutting steels has been displaced by the high quality and speed provided by fiber laser systems. High-definition plasmas are fast and efficient and can be considered for many projects involving cutting of steel, in particular those where post surface finishing will occur.

Another benefit of plasma cutting is that it can be used in the field to trim and cut steel. Using a straight edge or a template, plasma cutting large holes or shapes in the field is a viable option when trimming thin steel sheet to fit a particular application.

TORCH CUTTING

Torch cutting of steel is performed on thick steel. Cutting of beams, pipes, and tubes as well as thick plate with an acetylene torch is a common rough cut method. This method of cutting is used to trim steel for welding or for quick disposal. Figure 6.6 shows a steel plate that was torch cut and torch carved to produce a remarkable surface.

Torch cutting is fast but not particularly accurate. The edge produced needs to be addressed to clean up slag left behind.

WATERJET CUTTING

Cutting steel with a waterjet is another method to produce custom shapes and openings in a steel sheet or plate. Waterjet cutting pushes a tight stream of water at very high pressure. Accompanying the water is tiny fragments of garnet. The garnet in the water cuts through the sheet or plate.

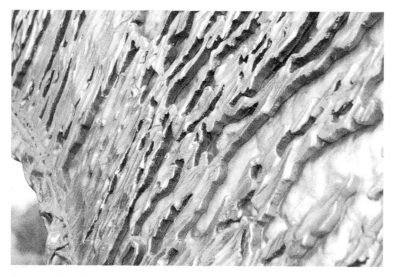

FIGURE 6.6   Torch-cut and torch-carved steel surface by the artist Rielly Hoffman.
Source: artist Rielly Hoffman.

Waterjets are extremely powerful and cut very thick plates. They are slower than laser cutting. The edge produced has a slightly frosted appearance. One issue on steel when the jet first starts to pierce the metal is that the water goes laterally across the sheet. This can make a frosted appearance if not controlled, and it will spray moisture across the sheet, which can lead to the development of rust. Plastic protection on the surface will help reduce the amount of surface hit with water, but the edge may still get moistened and the water can become trapped below the plastic.

Figure 6.7 shows weathering steel plates cut to emulate trees. These were CNC cut on a large-format waterjet machine and then weathered to the characteristic dark brown.

FIGURE 6.7   Waterjet-cut tree forms.
Source: Designed by Young and Dring Landscape Architects.

PUNCHING

Punch presses are a common form of cutting and perforating steels. They possess the power to pierce the metal. Die clearances need to be established for the particular thickness of the steel. Dies should be clean and polished and designed to pass clear through the sheet. Minimum hole sizes pieced through steel should be two times the thickness of the metal. For instance, if you plan to punch 2 mm thickness steel, the minimum hole size should be 4 mm in diameter. If the holes are not round holes, then use this relationship to extrapolate the minimum dimension to be punched.

When punching steel, it is important to take into consideration the work-hardening nature of the metal. As steel is pieced the edge around the hole is work hardened to a greater degree to that of the surrounding metal. This imparts differential stresses in the sheet and will require some stress relief or the sheet will warp excessively. Carbon steels work harden at lower rates than stainless steels. Punching steels can be performed at faster speeds with less worry about warpage of the sheet. Die clearance and planning for internal stress buildup in the sheet material is critical for successful punching processes. Figure 6.8 shows a preweathered steel wall that has a design induced by a CNC punch press operation. The sheets were flattened after the punched area was produced.

FIGURE 6.8   Custom-punched preweathered steel plates.
Source: Produced by ImageWall.

FORMING

Steel can be fabricated with conventional forming and shaping equipment. There are several significant differences, though, when fabricating with stainless steels. Steels will work harden as they undergo cold-forming operations. Steel forming will require more energy to overcome the stiffness and work hardening behavior of the metal. More tonnage is needed to form steel as compared to other metals of similar thickness.

For most simple bending operations, this matters little. But for stamping, spinning, and severe shaping, this will require a planned approach that may require interstitial annealing steps to reduce the internal stress. Inter-stage annealing involves heating the metal up to temperatures above 500 °C to remove residual stresses that have accumulated in the forming operations, this softens the metal and allows for further cold working without splitting the metal or damaging the dies.

ROLL FORMING

Roll forming is a common method of producing large panels of sheet metal from coils. The method involves a series of progressive rolls that gradually shape the metal ribbon into a linear form. Each successive matching set of dies alters the shape just slightly, which causes plastic deformation along the length of the metal surface.

Steel requires specialized equipment to achieve successfully formed panels using roll forming systems. Steel coil material entering the roll forming stations should be stress relieved to the greatest extent possible. It should be leveled in a tension leveling set of rollers that feed into the roll-forming station.

The calibration of the spacing of rolls and the matching dies must be set for steel. The dies must be made to account for the higher yield and the elastic behavior steel sheet will exhibit. As the metal enters the first set of forming dies, there is a stretching of slight lengthening that occurs. As the metal moves to the next station, further stretching happens. For steel, this is accompanied by cold working and hardening.

The best roll-forming operations have multiple stations spread out over several meters to reduce the stretching that has to occur and spread it out over the length of the station. Roll forming is a rapid forming process and is used to create metal siding panels from painted steel stock material as well as siding from weathering steel. Metal roofing, metal decking, and metal flooring are common articles produced by roll forming stations. Figure 6.9 shows examples of roll formed panels out of steel. The top two images are roll-formed weathering steel. The bottom is galvanized steel.

PRESS BRAKE FORMING

Brake forming is a common sheet metal cold forming process where a sheet or plate is inserted into a large forming machine called a press brake. A machined and hardened die is firmly fixed into

FIGURE 6.9   Roll-formed steel panels.
Source: L. William Zahner.

an adjustable holder. The tool or punch is mounted to the ram. The ram is powered by massive hydraulic cylinders that bring the punch down onto the sheet or plate resting on the die. The sheet or plate is formed to a given angle as the pressure is applied through the punch and through the die points.

Press brakes form material in a linear, straight line across a sheet or part. The bend can be 180° all the way back on itself or any angle between for the mild carbon steels. More power and force are needed, however, to shape thicker steels and the HSLA steels that make up the weathering steels (Figure 6.10).

A press brake can be fitted with special radius dies to induce straight, ruled curves into a sheet or plate. This allows for bull nose shapes and custom contours to be accurately produced in a press brake with the proper tonnage. In expert hands, a press brake is a highly versatile piece of forming equipment.

FIGURE 6.10   Press brake formed 3 and 4 mm thick weathering steel panels.
Source: L. William Zahner.

OTHER COLD-FORMING PROCESSES

Plate rolls are used for fabricating cylindrical or conic shapes. Plate rolls are a series of rolls set to interact in such a way they curve the plate by differential stretching of the surface. Steels are well suited for shaping with plate rolls, sometimes referred to as pyramid rolls due to the way three rolls are arrange with two lower and one upper roll. Figure 6.11 shows a typical setup used to roll large thick plates of steel.

The metal plate is fed between two rolls and pushed upward by the third roll in such a way they outer surface is stretched slightly in relation to the inner, or upward facing surface. Usually the process is incremental, meaning the slightly shaped plate is fed back through the rolls as additional shape is developed. Column covers, wall sections and other curved forms are cold formed in this manner. Figure 6.12 in the upper image shows a series of ribbons of carbon steel curved and welded, column covers made from 3 mm thick weathering steel.

FIGURE 6.11   Image of plate rolls.
Source: Courtesy of Taylor Forge Engineered Systems.

Steel beams, tees, and angle shapes can be curved using large hydraulic powered rolls that, similar to plate rolls, stretch one surface while restricting the other surface. These are usually cold processes that require significant pressure to shape the forms.

Circular cross-section pipe and tube are curved in a similar manner where surfaces are restricted while other surfaces undergo stretching. Figure 6.13 shows a decorative guardrail made of steel tubing and steel plate.

SPRINGBACK

Steel press brake forming requires understanding a characteristic of steels known as *springback*, the term used to describe when a material undergoes incomplete plastic deformation. It is a change in strain produced by what is known as elastic recovery. The higher the yield strength of a material, the greater the elastic recovery or springback. Springback is a dimensional change that occurs on formed parts when the pressure of the forming tool is released. Springback is a function of the materials' strength. See Figure 6.14.

All steel forming experiences some level of springback, depending on the ratio of bend radius and material thickness. There can be differences in strength within the sheet itself, creating fluctuations in springback. Alloys of steel with high-tensile strength such as the HSLA alloys have greater springback tendencies. It takes more power and force to achieve the same brake form on

FIGURE 6.12   Upper image is a curved ribbon of sheet metal designed by the artist, Jesse Small. The lower image shows weathering steel plates curved to form column surrounds.
Source: designed by the artist, Jesse Small.

the HSLA weathering steels as it does on low-carbon steel and an allowance for compensating for elastic recovery.

Brake forming a 90° leg in a thin plate may require overbraking a few degrees to arrive at a true 90° due to steels strength and resilience. In practice, it is a trial-and-error process to determine and gain an understanding of minor material differences. There are formulas that assist in approximation, but the typical practice involves first choosing a smaller bend radius and then advancing from there to a greater radius. The formed angle radius will be slightly smaller than the actual radius

FIGURE 6.13   Decorative steel art made from formed curved tube and plate.
Source: L. William Zahner.

achieved when the pressure is released. One can also bottom out the die or coin the die, but this can wear on the die material. It requires more tonnage and is not commonly used.

Air forming coupled with modern compensation equipment and sensors can overcome springback in the material. Air forming uses the two edges of the die with the single edge of the tool. Depending on the material thickness or inside radius, the springback will change. This can be adjusted by changing the die used.

HSLA weathering steel – 90° bend

Bend radius	1 thickness	6 thicknesses	20 thicknesses
Over bend in degrees	2°	4°	15°

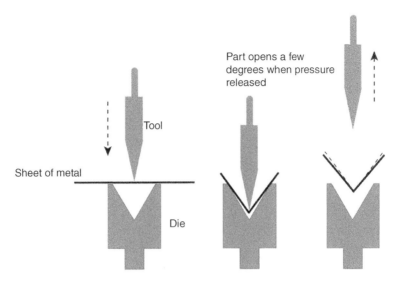

FIGURE 6.14   Depiction of what occurs in springback.

FIGURE 6.15   V-dies used on forming steel.
Source: L. William Zahner.

Air forming overcomes springback in the sheet metal brake form by pressing the metal down into the opening of the die. Less press tonnage is required to air form. Dies used in press brake forming are called V-dies (Figure 6.15). They come in various angles to compensate for springback. The sides of the die are available in various angles. The angle is designed to compensate for springback by pushing the metal toward the downward moving punch. V-dies are available at widths measured from tip of die to tip of die.

As material thicknesses increase or as the angle required increases, a die with a smaller angle may be needed to push the material toward the punch. The tool and die must be matched to give the needed clearance. The die angle must be greater than or equal to the angle on the tool or you may damage the tool.

90° die	Die opening range of 4 mm to 12 mm
88° die	Die opening range of 12 mm to 25 mm
85° die	Die opening range of 25 mm to 37 mm
78° die	Die opening range of 37 mm to 50 mm
73° die	Die opening range of 50 mm to 63 mm

During press brake forming, the metal toward the outer side of the bend goes into tension while on the inside of the bend, the metal is in compression. Metal along the edge work hardens more than metal away from the edge. To hold its shape, the metal must undergo plastic deformation, which means it will not return to its original shape with the loading is released. The metal strain hardens along the bend.

Bending or brake-forming steel will push the bend to a slightly larger radius than other materials due to the high strength and work hardening that occurs with steel. The bend radius, the radius of the curvature of the inside or concave side of the bend, for steels is larger than for softer, lower strength metals. When the metal is formed, the outer fibers, those on the outer side of the bend, are stretched, and if this is excessive, a crack can form at the bend.

Marring around the bend can occur from pressure of the steel to the die – in particular, on weathering steel. If the metal is preweathered before forming, the pressure needed to form the bend will mar the surface and create a line running parallel on either side of the bend. Figure 6.16 shows marring just off the corner of a folded thick plate. Note also the puckering that can occur at the corner when folding thick plate.

V-CUTTING

Steel can be V-cut. *V-cutting* is a term used to describe the physical removal of some of the metal at the point of a bend in order to allow for a tighter radius bend. V-cutting is a form of machining. The metal that is removed leaves a V-shaped cavity on the sheet of material before it has been formed. Specially designed equipment is used to remove the sliver of metal. A special cutting tool and tool holder are needed as well as a vacuum table or sufficient clamps to hold the steel sheet or plate firmly in position as the tool removes a thin sliver of metal. The force can be significant as the sliver of steel is carved from the back surface of the sheet. Figure 6.17 shows a darkened steel panel with a V-cut edge.

Steel is an excellent material to V-cut because of its nongalling characteristic.

Properly designed tools will not add excessive heat into the sheet, eliminating the need for coolants as the curl of steel is removed from the reverse side. V-cut steel, when folded, produces a clean, smooth edge. If the tool is loose, chatter will be apparent on the fold.

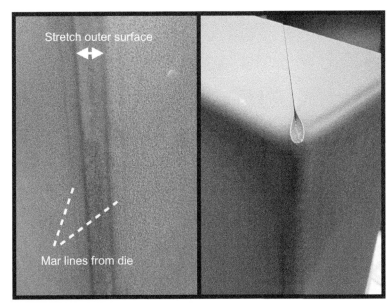

FIGURE 6.16 Marring caused by die pressure around the bend of thick plate. Note also the puckering at the edge due to pressure on the metal.
Source: L. William Zahner.

One interesting behavior of V-cutting is the alteration of stress induced at the fold. As stated earlier, brake forming places the outer surface of the bend in tension while the inner surface is put into compression. When the edge to be formed is first V-cut, then brake formed, there is a significant reduction in the tension induced stress. Compression is eliminated by the removal of the material. The effect is no contribution to unbalanced stress that might exist in the sheet as it is folded 90°. The removal of metal by V-cutting limits the compressive stress and allows for a much sharper bend.

V-cutting, however, does weaken the corner bend and with steels corrosion at the corner can lead to further weakening. It is not recommended to V-cut weathering steel, nor is it recommended to V-cut any steel used out of doors or in circumstances where corrosion may occur. If the V-cut is too deep or if the metal sheet is not held flat to the cutting table and a slight wave is cut into the back surface, this could lead to premature cracking.

HOT FORMING

Hot forming is usually performed on plate, bar, pipe, or tube. Hot forming steel requires close control of the heat. Usually, the steel is heated by means of induction. This is not a common practice in facilities that work to create steel parts for art and architecture.

Steels are readily shaped by hot-forming operations such as rolling, extruding, and forging. Curved shapes in heavy plate require hot-forming processes to enable the shaping, as shown in

FIGURE 6.17   V-cut corner on blackened carbon steel.
Source: L. William Zahner.

Figure 6.18. Cold forming these thick steel shapes requires immense pressure, and work hardens the metal, making any further forming difficult. Heating the steel form allows shaping to occur.

FORGING

Forging steel involves heating the metal and applying force rapidly to the surface. The challenges with steel are several. Malleable steels are often formed into plates and shallow shapes by cold and hot forging. Each subsequent impact to steel requires more energy as the shape work hardens are resists forming.

In art and architecture, forged steel parts are generally limited to hardware and small fixtures. Other industries incorporate forging of steel into their fabrication processes. Oil and gas, transportation, and aerospace will use forging and hot forming to produce intricate shapes. Figure 6.19 shows

FIGURE 6.18   Hot forming a steel plate, 55 mm ($2\,^3/_{16}$ in.) thickness.
Source: Image courtesy of Taylor-Forge Engineered Systems.

large forged pipe sections undergoing heat treatment after forming. Note the large-diameter forged extensions coming perpendicular to the pipe form.

MACHINING

Steel alloys can be machined with the right equipment, tooling design, and speed rate. The tools wear quickly with the HSLA alloys. The equipment used to machine steel must be robust and rigid. Chatter and vibration should be eliminated. The tools used to cut the steel must be sharp and designed to create small chips rather than long spirals.

FIGURE 6.19   Large forged sections undergoing heat treatment.
Source: Courtesy of Taylor Forge Engineered Systems.

The chips will take the heat away. Coolant can be used to assist in moving the chips away or an air blast will work as well. The tools should be tungsten carbide tipped or titanium aluminum nitride coated. High-speed steel with cemented carbide will also work well for machining the carbon steels. Figure 6.20 shows a steel plate that has been machined to form a tapered wedge shape. The surface is highly reflective as the cutting tool action creates a polished surface where the steel is removed.

Steel alloys best suited for machining have a higher content of lead and sulfur. Alloys such 11L37 or 12L14, for example, are excellent steel alloys designed for machining. The "L" stands for lead. Alloy 11L37 contains 0.15–0.35% lead as well as trace amounts of sulfur. Alloy 12L14, is also covered by UNS G12144, has 0.15–0.35% lead and 0.26–0.35% sulfur.

STAMPING

Steel is an excellent metal for stamping. Stamping is a general forming term that covers forming work performed on sheet metal. This could include brake forming, stretch forming, and drawing. Essentially, the metal undergoes plastic deformation as it shapes from pressure into a die or

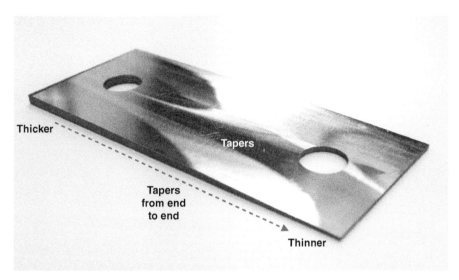

FIGURE 6.20   Machined steel plate to create a wedge form.
Source: L. William Zahner.

is restricted and pressed into a void. Typically, the edges are constrained and the metal plastically deforms under the pressure.

This is a rapid forming process with relatively high tooling costs. The metal work hardens and stiffens as the process unfolds. Grain structure and temper are important when the surface finish is important.

The transportation industry utilizes drawing and large stamping presses to create large steel body parts. The steel used is drawing steel quality (DS). This steel is specially produced to more exacting elemental constituents. The grain size is more refined to reduce directional tendencies that can have an effect on the forming behavior. There are other steels, deep drawing steel (DDS) and extra deep drawing steels (EDDSs). These are special carbon steels for specific operations. These have been processed to further improve the behavior of the forming process.

Steel manufacturing for forming operations has come a long way since the days of wrought iron and the construction of the Eiffel Tower in 1887. There were formed parts and assembled trusses but few stamped assemblies. At around the same time, there was a Belgium engineer named Albert Marie Danly. His firm was the Forges d' Aiseau. He invented a system of stamped iron or steel plates that, when assembled together, made a structural panel. He would incorporate iron bars at the joints to stiffen them. Sometimes the plates were galvanized; others were painted. Figure 6.21 shows the walls of the Edificio Metálico, the common name given to the Buenaventura Corrales Elementary School in San José, Costa Rica.

This amazing building has stamped and assembled steel walls, some with no interior structure. It was designed by the Belgium architect Charles Thirion. The accuracy of the stamped panels and the corrosion resistance of the metal are a testament to the durability of the material and the engineering of the firm that made them. There are over 1000 tons of iron. Fabricated in 1892 and shipped over to Costa Rica, the school was completed in 1896.

FIGURE 6.21 Edificio Metallico. Designed by Charles Thirion and manufactured by Forges d' Aiseau. *Source:* By Shutterstock.

There were other buildings around the world fabricated using what was known as the Danly system. Many were in tropical regions, like the Chalet Bosque in Belém, Brazil, or the Iron House in Maputo, Mozambique. This was the first of the prefabricated building. It could be made in Belgium and shipped anywhere in the world.

SOLDERING AND BRAZING

Soldering and brazing are similar processes that use an additive metal to join two metals together. In both processes, molten metal is applied to the joint, allowed to solidify and join the metals at the joint. This creates a metallurgical bond between the additive metal, the filler, and the surfaces of the metals being joined. This metallurgical bond is only as strong as the filler metal used.

Unlike welding, soldering and brazing processes are lower-temperature processes that have the ability to be reversed. You can add heat energy, melt the joint, and separate the two metals. The distinction between soldering and brazing lies in the temperature used to melt the filler metal and join the parts. The defining temperature is 450 °C.

$$\text{SOLDERING} < 450°C \ (840°F) > \text{BRAZING}$$

The success of soldering and brazing processes depends on the "wetting" ability of the filler metal. Wetting refers to the ability of the filler metal to be drawn into the joint sufficiently to make a good metallurgic bond. Fluxes are used to facilitate wetting behavior. A flux is a compound that removes the oxide on the base metal. Typically, a flux is a chemical compound, often acidic that when applied degrades the oxide by weakening the bond to the base metal. The flux penetrates the oxide though natural porosity of the oxide layer or tiny fissures or imperfections in the oxide layer. The flux remains on the surface during the process of joining and retards the development of a new oxide layer. They can also move the heat and draw the molten metal by reducing surface tension. This allows the molten metal to join to the base metal and displace the flux.

Fluxes by their nature are corrosive, and for steel it is very important to remove any excess flux from the joint after it has been soldered and brazed. Residual flux will lead to corrosion at the joint. The steel surface will rapidly show corrosion around the region where the flux is present.

Both soldering and brazing involve heating the base metal and the filler metal. The process introduces lower heat that welding processes; however, what often occurs is the molten metal expands the region around the joint, then when this solidifies, the distortion is held and permanently fixed into the part. As the molten metal cools, it pulls the thin steel surface toward it and creates a convex indentation on the reverse side. It is very difficult to arrive at good, distortion free joints when soldering or brazing.

Brazing zinc coated steels can be achieved but the zinc will make the joint weak. It is best to remove the galvanized where the brazing is to take place.

Other metals can be brazed to the steel. Copper and stainless steel can be joined to steel using the brazing process.

Thin steel sheets can be seamed together and soldered. Flux is required to remove the oxide layer from the surface of the area being joined. Flux allows the melted metal solder, usually tin–lead alloys, to wet the joint and flow into the seam, bonding the metal surfaces. Zinc chloride fluxes are still commonly used to remove the oxide from the steel, but it is critical to removal all excess flux after the solder joint has been accomplished. For overlapping seams this can be difficult if not impossible. Excess flux will corrode the steel in a matter of a few days. So, flushing the joint and removing all flux is critical. Solder used on steel can be any of the lead-tin, tin-antimony solders.

Steel has poor thermal conductance properties as compared to other metals such as copper or tin. When heated to soldering temperatures, usually near 205 °C, the heat is concentrated and will build quicker than soldering copper sheet.

Brazing is a higher-temperature process used to join two metals together. More common with copper alloys, brazing of steels is possible but controlling the heat induced warpage is challenging when brazing flat surfaces. Tubes and pipes are more effectively brazed due to their inherent section.

There are three categories of brazing used on steel, air with a flux, brazing is a reducing atmosphere and brazing in a vacuum. For all categories the area being joined must be thoroughly cleaned. Brazing temperatures are in the range of 600–700 °C. Metals used as the brazing fillers are alloys of silver, nickel, and copper. These have higher melting points than the tin–lead solders. Fluxes are needed to achieve flow into the braze joint.

WELDING

Welding of steel is in common practice for wrought and cast forms of steels. All the welding processes can be used on the steels. Welding is an ideal method of joining the low carbon steels. The welds that form are generally superior in strength over the base low-carbon steels; however, not so for many of the medium and higher content carbon steels. The welds can lack toughness due to the formation of martensite in the heat-affected regions.

Carbon steel melts around 1540 °C (2800 °F), which is a significant amount of energy. The electrical resistance of steel is also low. Carbon steel has a resistance of 12 $\mu\Omega$ at room temperature. As the temperature of the metal increases to around 890 °C, carbon steel's electrical resistance increases to around 125 $\mu\Omega$.

Steel has a thermal conductivity of around 50, which is higher than for stainless steels but considerably lower than aluminum and copper. Reference Table 6.2. Heat does not dissipate as rapidly away from the weld. A process of skip welding and chill bars are often introduced to aid in steel welding practices by reduce heat concentrations and by allowing the heat to dissipate to the surrounding metal.

Steels have a coefficient of thermal expansion half that of aluminum. Reference Table 6.2. Because of this, there is not a lot of warping from the heat of welding. Using techniques to remove the heat artificially with chill bars will reduce the warping that can occur.

TABLE 6.2   Coefficients of thermal conductivity and linear thermal expansion.

Metal	Coefficient of thermal conductivity (W/w °K)	Linear coefficient of thermal expansion (10^{-6} m/m °K)
Aluminum	240	22
Copper	400	16
Iron	90	12
Steel	50	11
Stainless steel	14	17

There are three goals to welding steel:

1. Achieve a strong connection.
2. Reduce warping.
3. Resist cracking from stress as the metal cools.

There are several processes of welding steel, and each has a set of criteria that go beyond the extent covered here. Welding processes add metal to the joint or they fuse the joint together without adding metal. When metal is added in the welding process, it called a filler metal.

Process improvements, use of robotics, and new heating techniques such as induction are changing the way steel is welded. The point of welding is to join one surface to the next and arrive at a system with good strength and equal mechanical properties as the overall body. There is resistance welding that is a forge-welding process in which pressure is applied as the heat of melting is achieved. This is also called *spot welding*.

Fusion Welding Processes Used on Steels

Fusion welding is a process where the metal is melted and fused, or metal is added to fuse two sections together.

Oxyacetylene Welding

This is a common method used to weld steels. It involves oxygen under compression and fuel, usually acetylene. Acetylene is a hydrocarbon, C_2H_2, and is the byproduct of the production of ethylene. Both gases are fed to a torch where the flame is adjusted by means of regulators to produce the necessary heat. Prior to the development of the electric welding processes, this was the most common method used to join steel. There are a number of safety systems required for safe use of welding gases, but the process of welding with oxyacetylene is one of the simplest.

Thick or thin steels can be welded. There are numerous torch sizes and set ups to facilitate the welding with this method. These torches can also be used to cut steels and carve the steel surfaces.

GTAW – Gas Tungsten Arc Welding

Gas tungsten arc welding (GTAW) is also known as TIG welding. In this process, a high-energy arc is created between a tungsten electrode, which is not consumed in the welding process. This process can produce a weld that requires very little cleanup. A major benefit to this welding process is lower heat input into the steel joint, which leads to lower oxide and slag levels. GTAW can be performed in all angles and all positions. It is a single-side welding operation.

Metal wire of the appropriate alloy can be fed into the arc at the joint to be welding. The wire is fed though a torch, which also distributes a shield of inert gas around the area of weld. Metal can be added manually by feeding into weld zone. Metal can also be fed into the weld zone automatically

FIGURE 6.22   Pipe assemblies welded, then galvanized after assembly.
Source: L. William Zahner.

using a timed feed system. GTAW is manual, semi-automated, and can be fully automated using robotic welding systems. GTAW is suitable for welding thin steels between 0.3 to 3 mm in thickness. Figure 6.22 shows several large pipe assemblies welding using GTAW methods.

GMAW – Gas Metal Arc Welding

Gas metal arc welding (GMAW) is also known as MIG welding. It is similar for all intents and purposes to GTAW with the exception that the wire is continuously fed and creates the arc without the tungsten. The wire electrode is consumed as the welding proceeds. The filler metal is automatically fed into the weld zone. GMAW is a very fast and efficient process that can be semi-automated or automated. Systems can be set to weld both thin and thick steels. The most optimum steel thicknesses for GMAW are between 2 and 10 mm. A shielding gas is used to keep the weld clean and free from oxygen and nitrogen contamination from the surrounding air. No flux is needed. The process produces very low oxide and slag. The GMAW can be undertaken in all positions and both sides of a part.

SMAW – Shielded Metal Arc Welding

Shielded metal arc welding (SMAW) is also known as stick welding. In this process, a high-energy arc is created (struck) between a current carrying electrode and the metal to be fused. The electrode feeds steel wire to the joint to be welded. The wire has a coating of flux that melts in the high temperature of the energy arc and provides protection and cleans the area directly at the weld. The high heat energy melts the metal being fed into the joint and the metal being welded. This is a common manual approach. It is not easily set for automation.

FCW – Flux Core Wire Welding

This wire-welding process uses a hollow wire containing a high temperature flux. It is also known as flux core arc welding, FCAW. The wire is passed through a welding apparatus and enters the high energy arc. No external inert gas is required because the flux reacts to develop a shielding envelope around the weld. The benefit occurs where instances of welding in wind can blow the shield gas away; here, the shield gas is produced on the target weld as the weld in laid down. Cleaning down the weld is more challenging due to the introduction of the flux at the point of weld.

SAW – Submerged Arc Welding

In this welding process, a steel alloy wire electrode carries a current and creates the high-energy arc. Flux is added by a separate feed into the joint being welded. The arc is created within the flux cover. This is a high productivity, automated process used on mainly flat surfaces. Heat input is high when using this process, so it is more suited to the heavier thicknesses, 10 mm or thicker.

PAW – Plasma Arc Welding

Plasma arc welding (PAW) uses a special welding torch that constricts the high-energy arc by means of a special nozzle. This increases the temperature and concentrates the heat. Like the GTAW process, bare metal wire is added and consumed at the weld joint. Plasma jet is very high temperature around 28 000 °C (50 432 °F). Depending on the weld type, PAW can weld thin steel, less than 2 mm and thicker plate up to 25 mm.

EBW – Electron Beam Welding

This is a specialized process performed in a vacuum. The EBW process aims a dense stream of electrons at the area to be welded. The high energy of the electrons is converted to heat sufficient to melt the metal. Usually, the metal joint is designed so that this very fine stream of particles fuses the metal edges together without filler metal added. Thicker metal can be welded to thinner metal. The EBW system produces low distortion in the welded parts. You can very accurately control the depth of the weld penetration and the heat is minimized. This produces very low distortion. This factory-automated process requires a high vacuum that will eliminate oxidation. This is not a common art and architecture welding process.

Laser Beam Welding

The process of welding steel with a laser focuses a fine high-energy beam at the point to be fused. The metal absorbs the energy at the exact point and melts from the tremendous heat that is developed. The main benefit is the reduced heat-affected zone. The laser can be set to provide the energy to just melt the metal. This reduces distortion and confines the area affected by the weld to a minimal zone. Properly welded joints with laser will require very little cleanup. The challenge is fixturing the parts to be welded. Laser welding requires CNC control or robotic control. The accuracy of precise fit-up and laser path layout cannot be ignored.

Laser welding the low-carbon alloys is very efficient and clean. Hardly any cleanup is necessary after the welding. Due to the small area of heat, the metal cools rapidly. Some of the elements in the alloy will come out of solution quicker and solidify, causing embrittlement around the weld.

Laser beam welding is a factory process that can be used on steels of varying thickness, depending on the power. Usually, CO_2 lasers are used for welding due to their higher power, although Nd:YAG can weld thinner sheets just as well. Powerful CO_2 lasers can weld up to 38 mm thick plates and are used to produce steel structural forms.

Capacitor Discharge Fusion Stud Welding

This is a method of attaching a stud to a steel surface by means of high-energy discharge. The capacitor discharge fuses small-diameter studs threaded and unthreaded to a thin sheet of steel. The stud is held against the surface as a capacitor is activated to release stored energy. The energy is sufficient to melt the face of the stud and the surface of the sheet material. The stud is pushed into the molten metal, which cools and fuses the two surfaces. It happens very rapidly, and distortion is reduced.

Arc Stud Welding

In the arc stud welding process, heavier steel plate is used as well as larger studs. A ceramic cup surrounds the base of the stud at the point of contact to the plate. The stud is raised just off the surface as energy is passed between the parts. The ceramic cup concentrates the heat to the area just around the weld and reduces weld splatter. When the base metal plate surface is molten and the surface at the end of the stud is molten, a spring-loaded ram shoves the stud into the molten metal. Solidification occurs rapidly, and the stud is set perpendicular to the surface.

Friction-Stir Welding

The friction-stir welding process was developed in Cambridge, England, in the early 1990s. The process involves a high-speed tungsten alloy tool spinning as it moves between two edges to be welded. Used extensively on aluminum, friction-stir welding has shown success in welding plates of low carbon steel together. The spinning tungsten tool generates heat from friction sufficient to melt the steel. As the tool passes between two edges it melts, the metal immediately around the spinning tungsten tool and the trailing metal resolidifies as it cools, effectively joining the plates together.

When used on steels, the tooling wears down rapidly, thus making the process less viable for joining steel together. This welding technique is not in regular use by metal fabrication companies involved with art and architecture. See Table 6.3.

TABLE 6.3   Comparative weld processes used on steel.

Welding process	Advantages/Disadvantages
Oxy-acetylene gas welding	Simple, versatile method. Manual. Can be used on thick and thin steels.
Gas tungsten arc welding	Low heat, thin metals, Automation and manual. Requires very accurate fixturing. Single side.
Gas metal arc welding	Fast and efficient. Automation and manual. No flux needed. Low oxide and little slag buildup. Thicker sheet. Weld cleanup more than GTAW.
Shielded metal arc welding	Rapid and versatile process. Automation of process is difficult. Slag and flux cleanup required. Thin and thick steel.
Flux core wire welding	No shielding gas is needed. Can be automated. Cleaning difficult. Higher temperature.
Submerged arc welding	No shielding gas is needed. High productivity. High heat buildup. Creates more ferrite in weld. Thicker plate only.
Plasma arc welding	Good weld quality. Filler metal is added. Automated system. Thin and thick steel. Generates high heat.
Electron beam welding	Excellent weld quality. Expensive systems. In factory process. Very accurate and low heat input. Both thin and thick steel can be welded. Requires the weld to occur in a vacuum.
Laser beam welding	Excellent weld quality. Expensive systems. In factory process. Automated. Can weld thick or thin sections depending on power output. Minor cleanup.
Capacitor discharge welding	Thin plate stud welding. Show through to face side can be challenging.
Arc stud welding	Heavy plate stud welding. Not suitable for thin sheet.
Friction-stir welding	Fuses plates from heat of friction. Limited use.

THE STEPS IN WELDING

For successful welding of steel, there are a series of distinct steps that, when planned and followed, will deliver the results intended.

All the processes used to weld steel require planning in order to be successful. The area where welding is to take place must be clean and free of contamination from other metals, sand, and moisture. Fixturing and support for holding the material to be welded must be rigid and clean. Chill bars, protective atmospheres, and proper weld techniques must be followed to aid in cooling and achieving a finish weld that meets the project requirements. Elaborate forms and shapes can be assembled easily and accurately using steel.

Joint Preparation

Beveling joints in thick steel allow for more passes with less heat buildup to create the joint. It is critical to ensure oils and grease are not on the edge. These can contaminate the weld. Waterjet cut edges need to be treated before welding to clean the garnet fragments that can be left embedded in the edge.

Filler Metal

Filler metal is the added metal introduced to the weld. The correct filler metal is important. When welding the steel alloys, it is recommended to use similar or matching alloy to the base metal. By matching, you want to match the strength of the surrounding metal. Matching strength of filler metals to that of the base metal is not defined in American Welding Society (AWS), but it implies the filler metal tensile strength should be equal to or greater than the tensile strength of the base metal. The standards center on the type of joint and the load applied to the joint. For example, if the base metal has a yield of 65 ksi (450 MPa) the filler E70 has a minimum 70 ksi (480 MPa).

There are several classifications defined by the AWS. These classifications are briefly described in Table 6.4.

Heat Input During Welding

On thick steel, multiple passes are necessary to lay down the correct amount of weld. Carbon steel alloys require specific heat ranges.

Post-Weld Treatment and Cleanup

Post-weld treatments of steel assemblies removes any residual stress that may be between the welded joint and the base metal as the weld cools. This is particularly important for thicker sections that may have developed stresses that exceed allowable design loads.

TABLE 6.4   AWS welding classifications.

Classification	Elements added	Welding specifics
A	Carbon and molybdenum	SMAW, FCW
B	Chromium and molybdenum	SMAW, FCW, GMAW
C/Ni	Nickel	Welding HSLA steels
D	Manganese and molybdenum	Welding HSLA steels
G	Special case. Alloy not defined	
K	Manganese, nickel, and molybdenum	GMAW, FCW
M	Military classifications	
W	Copper	Weathering steel

In the post-weld treatment, the joint assembly is heated to a specific temperature and held at that temperature for a period of time in order to reduce the level of stress that has developed in the welded assembly.

Most art and architectural welded subassemblies will not experience stresses that exceed the design loads so subsequent post treatments may involve more cosmetic cleanup than stress relieving the welded assembly. This may involve removal of oxides, filling cracks and craters in the weld bead. Eliminating the arc strikes and weld splatter are two common post treatments. Good welding skill and practice should eliminate the visual imperfections that can ruin the appearance of a welded assembly.

Distortion

An additional aesthetic issue is weld distortion. More common in the reflective stainless steels, weld distortion, particularly on thin materials, is a challenge faced with steel. On thick sections, the cooling behavior can rotate the surface around the joint. Distortion can be introduced by the heat input of the welding process, uneven and rapid cooling of the assembly, or imparted by grinding and weld-reduction techniques. If the elements being joined by welding are curved and arching sections, cleaning and grinding down of the welds can be a significant challenge. With straight, flat sections the cleaning down of the welds can be staged using tapes and block sanding. Not so with curved surfaces. These will require very careful removal using gauges and curved protection barriers. If the assembly is to be galvanized, this will not be as large an issue however, if the surface is painted, the grinding marks and uneven surface can be accentuated by the gloss of the paint.

Joint design, welding technique, preparation, grinding, and finishing techniques are important considerations to be worked out for a successful result. Prototype construction, to prove out technique, is a good path to begin and learn from when approaching complex fabrication using steel. This will establish the best course of action necessary to achieve the desired results. Designing devices to facilitate removing the heat generated during the welding process can be one of the critical steps in achieving a distortion-free surface.

Steps to Reduce Distortion in Welded Surfaces
1. Straighten and smooth plates.
2. Prepare and clean edges to be welded.
3. Test and fit up joints before weld application.
4. Add shape to the surface if possible before welding.
5. Locate weld at edges.
6. Reduce tack weld size.
7. Continuously check fillet welds using fillet gauge.
8. Use automated systems.
9. Incorporate pattern weld techniques to reduce heat buildup.
10. Use heat sinks in welding process.
11. Step grind and use gauges such as tapes to prevent overgrinding.
12. Use several steps in sanding, starting with larger grit down to smallest grit.
13. Test process on prototype assemblies.

WELDING OF WEATHERING STEELS

Weathering steel is easily welded using common welding processes such as submerged arc, shielded metal-arc, gas metal-arc, and even flux cored arc welding.

Low-hydrogen electrodes are recommended for welding weathering steel. To give good color match and strength, E70, E80, E90, E100, or E110 electrodes followed by W electrodes of close to the alloy of the metal to give color match are recommended on large plate fabrications. These electrode types can provide the strength inherent with HSLA weathering steels.

For simple, single-pass welds, where structural concerns are minimal, the base metal will dissolve into the weld. The weld material will pick up the constituents in these corrosion-resistant alloys, and essentially the weld color and consistency will be very close to that of the base metal. Therefore, carbon steel filler metals can be used.

On thicker sections, when the weld requires multiple passes, there will be less dissolution of base material into the weld material. Low-alloy-type filler metals should be used. Filler metals should have a minimum nickel content of 1% such as the Ni1 filler metals or the W filler metals composed of copper, nickel, and chromium. The W filler metals will give good color match on the surface.

For multiple pass welds, the lower welds could be made with the carbon steel filler metals and the top weld passes could be made with the more expensive W filler metals for color match. Refer to Figure 6.23.

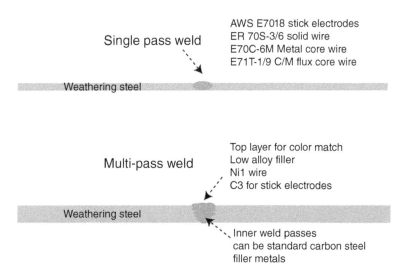

FIGURE 6.23 Welding thin versus thick weathering steel.

BOLTS

Weathering steel bolts and nuts are available for attaching weathering steel plates. These are very high strength and give the appearance characteristic of the metal they are joining. Galvanized steel fasteners can also be used, but you will get some staining below the fastener as the zinc works to protect the weathering steel.

Stainless steel screws and rivets work in thinner assemblies. They lack the color and painting them dark will help match the color of the darkening steel.

It is not advised to use plain steel fasteners, even if they are painted. The process of preweathering will cause the steel fastener to corrode, and without the benefit of the alloying constituents in weathering steel it will continue unrestricted.

Brass fasteners have been successfully used in weathering steel. The small area of brass used to join plates of weathering steel will have minor dissimilar metal concerns. The brass is soft compared to the steel, and the threads can be easily stripped but they tend to darken and match the color of the weathered steel.

When assembling post-galvanized parts, it is important to ensure the holes edges are coated. Bolting of galvanized steel is common. Welding will damage the galvanized coating. You can touch up the welds with a zinc coating treatment, but the color tone and performance of the coating will not be the same as the balance of the galvanized surface. Figure 6.24 shows two sculptures made from welded assembles, galvanized, then shipped and assembled with bolts in the field.

FIGURE 6.24   Welded and bolted steel towers. Post galvanized.
Source: Designed by the artist Ron Fischer.

STUD WELDING ON WEATHERING STEEL

Fusion stud welding using steel, stainless steel, or copper studs will work on weathering steel. You have to remove the oxide around the stud interface first. The bond is excellent, but you should consider zinc paint around the weld interface.

CASTINGS

Iron and steel are both available as a cast form. Cast irons contain 2–4% carbon and 0.5–3% silicon.

TABLE 6.5 List of cast iron types.

White cast iron	Fe$_3$C and pearlite
Gray cast iron	Graphite flakes
Malleable iron	Graphite nodules
Ductile iron	Graphite spheroids
Compacted graphite cast iron	Graphite reduced or removed in anneal

Types of Cast Iron

There are five main types of cast iron. See Table 6.5.

White Cast Iron

White cast iron is the beginning point for the malleable cast irons. When fractured the inner surface appears white. White cast iron is hard, but it fractures more easily when subjected to bending forces.

Gray Cast Iron

Gray cast iron has small flakes of graphite that induce ductility into the casting but at a cost of strength. When fractured, it has a dull gray color. This is the most commonly cast form of iron. It has decent compressive strength, good machinability, and good vibration dampening characteristics. See Figure 6.25.

Malleable Iron

The iron form is produced by heat treating white cast iron. Clumps of graphite develop, and this induces better ductility and good machinability. There are three subforms of malleable cast iron: blackheart malleable iron with 2.2–2.8% carbon, whiteheart, malleable iron with 2.8–3.4% carbon, and pearlite malleable iron when pearlite forms as the iron is annealed and cooled in air or oil to form pearlite.

Ductile Iron

Also known as spheroid iron casting, ductile iron is produced by adding magnesium or cesium, which causes the graphite to form into spheroids during solidification. This form of iron has good strength and ductility.

FIGURE 6.25   Artistic gray cast iron sculptural base in Seville, Spain.
Source: L. William Zahner.

Compacted Graphite Cast Irons

These cast irons contain spherical graphite that are interconnected and much more compact. They are called vermicular graphite due to their appearance. They have high strength. They contain a small amount of titanium to form the compacted graphite.

Difficulties in Casting

Cast iron and cast steel can possess flaws induced during the casting process. Common flaws include cold shuts, holes, cracks, and embrittlement. Cold shuts are a partial separation of the metal, like a crack or split that does not go all the way through. It can occur on large, thin surfaces or where two

TABLE 6.6   Potential Casting Flaws and Their Probable Causes

Flaw	Probable cause
Porosity	Trapped gas or outgassing as the iron solidifies
Shrinkage	Carbon or nitrogen content in melt
Flash	Mold not fitting up and metal projects at parting line
"Rock candy" fractures (conchoidal)	High nitrogen or aluminum in melt, causing the metal to separate along grain boundaries
Blow holes and pin holes	Trapped gas, similar to porosity
Lustrous carbon	Only visible at fracture; binders in mold enter into and affect the melt
Fillet veins	Binder in mold, causing a crevice to form
Misrun	Lack of fluidity in melt so metal does not flow to the edges
Cold shut	Similar to misrun but where large surfaces are not completely filled
Warp	Residual stresses develop during cooling
Buckle	The sand expands and prevents the mold from filling
Cat tail	Small sharp metal "strings" extending from the gates due to mold separation at these points
Cold shot	When small drops of metal solidify before the mold is fed with the melt
Slag inclusion	Reaction between the mold substances and the metal, causing a compound differing from the metal to form on the surface
Expansion scab	The mold expands and thin metal projections occur
Ramoff	Casting shows thickening along the mold parting line
Kish graphite inclusion	Course porosity on the surface, excessive carbon in melt
Scars	Waves or folds on the surface plan

or more pours meet up in the mold. Holes are caused by trapped gasses in the casting. Cracks can be generated from differential cooling or improper sand. Embrittlement can develop from graphite forming in the cast iron or steel. Table 6.6 lists a number of potential flaws and the probable cause of the flaw in casting.

Common Casting Flaws in Steel and Iron

The critical part is the final finish necessary to provide both the appearance and the durability required of most art forms. For many steel surfaces, the initial process involves blasting to remove the oxides and slag that are on the surface. Chemical pickling processes are also used, but they involve strong acids that require special handling and processing.

Blasting can be performed with various grits and grit sizes. Clean sand, steel shot, steel particles, garnet, and glass beads are all common media used on steel surfaces. They leave the surface with microscopic irregularities, often referred to as *tooth,* that receive additional coatings well.

Corrosion Characteristics

It rusts and it sinks in water.

... British Admiralty when ironclad warships were suggested

INTRODUCTION

For all the strength, ductility, and shaping potential that iron and steel offer, they give it all up in their lust for oxygen. Like the mythical Samson, iron's strength is sapped by its desire for combining with oxygen – it's Delilah.

Iron cannot resist oxygen. Which is a good thing for life on this planet. Iron in our blood system bonds with oxygen in our lungs in a complex hemeprotein molecule called hemoglobin and is responsible for the delivery of oxygen to parts of our body. Iron is what gives blood its red color.

There are no instances of native iron that is iron found in a near pure state, on Earth's surface. Some meteorites are of exceptional purity, but for all other iron, it is found in combination with other elements. Iron and oxygen combine in six common forms (see Table 7.1), but as many as 16 forms exist. If oxygen is present, iron will find a way to join with it.

The two elements, iron and oxygen, need an introduction of sorts, though—an enabler, something that helps set the stage that will empower the two elements to interact. Moisture is that enabler in the form of rain and condensation when changes in relative humidity of surrounding air while the steel surface temperature stays cool. Figure 7.1 depicts what often happens as the ambient temperature changes, the metal temperature lags.

It also helps if there are substances in the moisture, either collected from the air or off the surface of the iron, that will convert the moisture into an electrolyte by supplying ions to carry a current. There are enough particles from the surrounding air or elements within the steel alloy exposed on the surface that can diffuse into the water and form an electrolyte. Steel, of course, is the most

TABLE 7.1   Six common forms of iron oxide.

Compound name	Chemical formula	Color
Iron (II) oxide	FeO	Black
Iron (II) hydroxide	$Fe(OH)_2$	Green
Iron (II, III) oxide	Fe_3O_4	Black
Iron(III) oxide	Fe_2O_3	Reddish brown
Iron (III) hydroxide	$Fe(OH)_3$	Yellow orange
Iron (III) oxyhydroxide	$FeOOH$	Dark orange

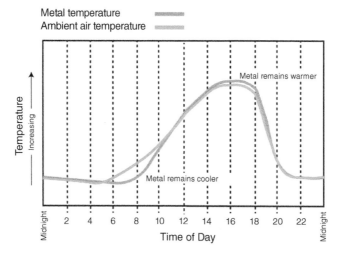

FIGURE 7.1   Ambient temperature and metal temperature.

common alloy of iron, containing very small amounts of carbon, less than 1%, along with other alloying elements.

Steel is less corrosion-resistant than pure iron because of the presence of other elements within the alloying mix. But steel is more corrosion-resistant than cast irons, which are also alloys of iron containing around 3% carbon.

With the presence of moisture, the steel surface will begin to set up small, localized galvanic cells. Small quantities of oxygen are all that is needed in the water to cause ferric ions to enter the solution. Which form of oxide the rust will take depends on the amount of ferric irons.

In the absence of moisture, the corrosion of the steel surface is very slow. Thus, in desert climates or in very cold climates below the freezing temperature, corrosion of steel and iron is greatly reduced. Without the moisture, the electrolyte will not develop. Corrosion of metals is very dependent on climatic conditions. Still, for steels, not a lot of moisture is necessary to cause corrosion cells on the surface to develop.

For example, Figure 7.2 shows an interior surface of blackened steel. Exposed to the relative humidity and temperatures that are common to an office setting, this corrosion formed over several years.

In most waters, dissolved oxygen is present. It is the dissolved oxygen that reacts with hydrogen absorbed on the surface of the steel at dispersed locations across the surface. The oxygen and

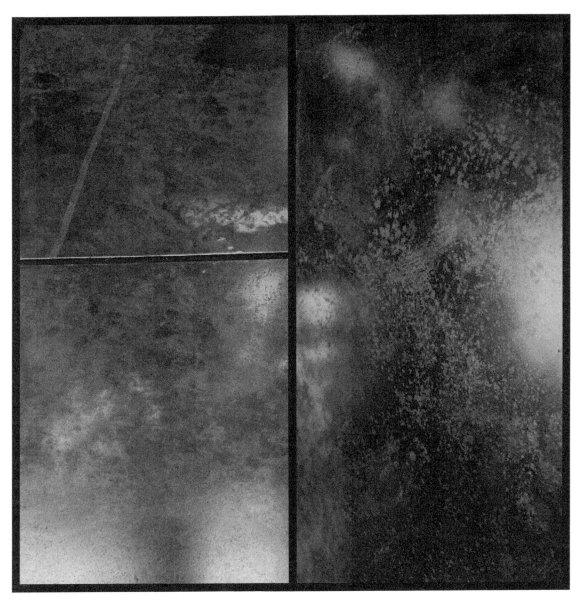

FIGURE 7.2   Image of a blackened steel wall with rust spotting.
Source: L. William Zahner.

hydrogen form hydroxide, (OH)⁻, which combine with iron atoms diffusing from the surface into the electrolyte. Corrosion will continue as long as the supply of oxygen is there.

The protective film that first develops on steels is actually more corrosion-resistant than the film that initially forms on copper, but it fails rather quickly. The atmospheric corrosion attacks the surface of the steel in narrow bands as opposed to copper, which undergoes a uniform formation of oxide.[1] Figure 7.3 shows corrosion on the surface of an A36 carbon steel plate that has been stored a short time out of doors.

As previously identified, there are as many as 16 forms of iron oxide and hydroxides. The most common iron oxide, we know as rust. Rust is ferric oxide or iron (III) oxide. Ferric oxide, Fe_2O_3, is similar to the mineral hematite, the main mineral ore source of iron for the steel industry. Hematite contains approximately 70% iron. Hematite, in fine granular form, is what gives jewelers rouge the red color, and its relative hardness enables the polishing of softer metal surfaces. Ferric oxide is nonmagnetic; it lacks the magnetic attraction, the ability we associate with the base metal iron and its main alloy steel.

The other common mineral form is magnetite, Fe_3O_4, also known as iron (II, III) oxide or as ferrous oxide. Magnetite contains 72% iron, and the crystal structure of this iron oxide is cubic causing

FIGURE 7.3   Images of steel plates corroding.
Source: L. William Zahner.

[1]Weissenreider, J. and C. Leygraf. In-situ Studies of the Initial Atmospheric Corrosion of Iron. pp. 127–138. In Outdoor Atmospheric Corrosion, Herbert E. Townsend (ed.), West Conshohocken, PA: American Society for Testing and Materials International, ASTM Stock Number: STP1421.

TABLE 7.2 Mineral forms of rust on steels.

Ferric oxide	Iron (III) oxide	Fe_2O_3	Hematite	Reddish brown
Ferrous oxide	Iron (II,III) oxide	Fe_3O_4	Magnetite	Black
Ferric oxyhydroxide	Iron (III) oxyhydroxide	$FeO \cdot OH$ $Fe(OH)_3$	Lepidocrocite	Yellow to orange

it to occasionally exhibit magnetism. The famous and mysterious *Lodestone* is magnetic magnetite. It gets its magnetism from the way the atoms group together and the way the electron pairs become aligned. Essentially, magnetite is composed of both Fe^{2+} and Fe^{3+} ions (II and III nomenclature). These two different charged ions produce an imbalance as the negatively charged electrons flow between the two types of iron creating a magnetic field (Table 7.2).

WHAT IS RUST?

Iron wants to return to its natural state of oxides and hydroxides, which are essentially different mineral forms. For art and architectural steel assemblies, corrosion is in the form of a general surface change that occurs when the steel object comes in contact with a moist environment.

The reaction is defined as follows:

$$4Fe + 6H_2O + 3O_2 \rightarrow 4Fe\,(OH)_3$$

This is the common corrosion of iron. The formula describes how iron present on the surface, along with water and oxygen react to form $Fe\,(OH)_3$, also known as iron (III) hydroxide or ferric hydroxide. The oxygen in the equation comes from the atmosphere and from dissolved oxygen in the moisture on the surface. Oxygen readily dissolves in water, particularly natural waters. Ferric hydroxide, $Fe\,(OH)_3$, the product in the formula, is insoluble and comes out of solution and deposits on the steel or flakes off. The ferric hydroxide is a reddish-brown colored rust.

Often forming with this oxide is $FeO\,(OH)$, or ferric oxyhydroxide. It is soluble and forms on steel when standing water is on the surface. When this occurs, it is considered hydrated, $FeO\,(OH) \cdot nH_2O$. The color is yellow or orange. Because it is insoluble it will run and drip from the surface as shown in Figure 7.4.

As the surface dries out, the reddish color we associate with rusted steel remains. This rust is ferric oxide, Fe_2O_3, also known as iron (III) oxide. It is the form of rust that develops over a surface during uniform corrosion.

The formula for this is:

$$2Fe(OH)_3 \rightarrow Fe_2O_3 + 3H_2O$$

Ferric oxide is insoluble and more adherent to the steel surface. When oxygen is in low supply, the ferric oxide can appear black in color. This black stain is very stable and adherent. Figure 7.5 shows the black form of this oxide, formed from heavy chloride attack.

FIGURE 7.4   Stain from corroding steel set over a copper roof.
Source: L. William Zahner.

Sometimes another form of the hydroxide forms, ferrous hydroxide, $Fe(OH)_2$, also known as iron (II) hydroxide. It differs slightly from its brother, ferric hydroxide and is a more unstable form, leading to development of other oxides as oxidation progresses.

The formula is as follows:

$$Fe + H_2O + \tfrac{1}{2} O_2 \rightarrow Fe(OH)_2$$

Ferrous hydroxide will often form on the surface in environments high in chlorides. The color of ferrous hydroxide has a greenish tint. Figure 7.6 shows green rust along with an orange color ferric oxyhydroxide. The black is iron (III) oxide where it was covered and devoid of oxygen.

These oxides, commonly referred to as rust, are porous and fractured because it takes up a larger volume than the base metal and swells as it grows and expands. That is why a corroding surface is

FIGURE 7.5   Black rust deposit on steel.
Source: L. William Zahner.

rough compared to the original smooth surface. The swelling, porous surface can act as a sponge, holding more water near the base metal.

The iron oxide that initially forms lacks cohesion and easily detaches and flakes off the surface, exposing more of the base metal to continue the oxidation process. It is this affinity to oxygen that limits the use of uncoated steels apart from what are referred to as weathering steels, which will be discussed further in this chapter.

This expansive growth of the oxide on exposed surfaces makes for an aesthetic issue as well as a structural issue. The development of the oxide can have dire consequences because it is not irreversible. This strong metal is rendered friable and weak. Figure 7.7 is a handrail exposed to salt-laden air. After a few years, the steel has expanded and become friable. What can be even worse is when the steel is encased in stone or concrete. The pressure developed by the expanding volume of iron oxide can break stone and spall concrete, as shown in Figure 7.8. This is known as rust burst or oxide

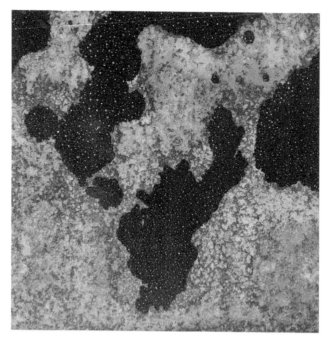

FIGURE 7.6   Green rust forming along with other oxides.
Source: L. William Zahner.

jacking. Iron oxide doubles the volume previously occupied by the original metal. This swelling can produce so much force that concrete or marble will crack. This irreversible, near-complete destruction is all because of iron's lust for oxygen.

CORROSION OF STEEL AND IRON

Corrosion of steel and iron is a degradation process in which there is a dissipation of energy in some form or other. Corrosion in natural waters and common exposures involves dissolution of iron into an electrolyte. The electrolyte can contain chlorides from salts or mildly acidic solutions formed from particles in the air. Dissolution is the result of electrochemical reactions on the surface of the metal. This type of corrosion causes the iron to combine with the oxygen in the water to form iron oxide. The ion of iron on the surface exchanges two electrons with hydrogen in the solution, leaving it with a 2+ charge. The formula is:

$$Fe + 2H^+ \rightarrow Fe^{2+} + H_2$$

Water is the most common natural oxidizing agent to start the process of forming iron oxide on the surface of steels and iron. Even humid atmospheres can cause iron oxide to form. Corrosion occurs when an anodic polarization is established and the iron atom, once in equilibrium, is

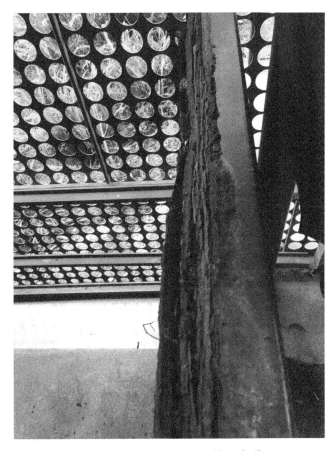

FIGURE 7.7   Severe corrosion of along the edge of a painted steel handrail.
Source: L. William Zahner.

FIGURE 7.8   Stone breakage from rust.
Source: L. William Zahner.

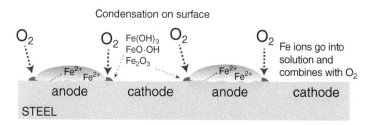

FIGURE 7.9   Rust formation on steel.

now out of balance as the electrons are exchanged with hydrogen present on the surface. Figure 7.9 depicts what is occurring as the steel surface begins to corrode. The removal of electrons is called the oxidation of iron. This leaves the iron as positive ions, Fe^{2+}. These ions combine with the oxygen in the moisture and form the compound ferrous hydroxide, $Fe(OH)_3$ often along with $FeO \cdot OH$, an oxyhydroxide. The ferrous hydroxide can be green, reddish-brown, or black. The oxyhydroxide is orange-yellow in color. A third oxide is often present, Fe_2O_3, which is the ferric oxide form. These three compounds make up what we refer to as rust.

When chlorine is present in moisture, it is always in its ionic form, Cl^-. This form increases the conductivity of the moisture, making it an electrolyte. Seawater contains around 3% chlorine, which is sufficient to make it conductive. It has been found that this percentage of chlorine is optimum for corrosion conditions on metals. Essentially, a higher chloride concentration does not change the corroding power of the seawater.

The chloride ion in moisture on steels is responsible for establishing the localized corrosion cells that rapidly develop. The steel surface with all its irregularities rapidly forms anodic and cathodic regions when moisture laden with chlorides forms on the surface. Condensation forming on the surface can collect and dissolve salts that have dried on the surface, creating this conductive electrolyte. The areas just under the electrolyte become anodic, while the areas outside the electrolyte are relatively cathodic. The iron ions dissolve into the electrolyte and seek out oxygen in the moisture and at the edges of the moisture.

When chlorides are present in the electrolyte, say from saltwater or from deicing salts on the surface, they are not bound to the iron but instead form an instability on the iron surface. This unstable iron, unstable because it has relinquished its electrons in the outer shell, combines with oxygen to form iron oxide. The chloride then moves to other iron on the surface to continue the reaction. In this way a small amount of chorine can do a lot of corrosive damage to the iron.

ATMOSPHERIC CORROSION – ELECTROLYTES ON THE SURFACE

Atmospheric corrosion is considered what a metal surface would experience in a moist environment in normal temperate conditions. Corrosion that a metal would experience is electrochemical in nature and highly dependent on the thin film of moisture that forms on a surface. This film is an electrolyte because it has absorbed various gasses in the atmosphere, airborne particles as well as substances that are on the surface of the metal object including corrosion products that begin to form.

The rate of corrosion a metal surface will experience depends on the nature of this electrolyte. Characteristics of the substances dissolved in the electrolyte and the thickness of the film are important parameters. Films formed by condensation are thick and develop over a few hours. As they form, they gather substances on the surface of the metal. Condensation is different than rainwater. Rainwater hits the surface and can actually remove substances from the surface, whereas condensation forms slowly and gathers, holding substances so they can dissolve and concentrate.

RATE OF CORROSION

- Thickness of electrolyte
- Ions in electrolyte
- Time electrolyte remains on the surface

Condensation of a smooth, flat metal surface occurs around 100% relative humidity. However, in small crevices or on concave surfaces, the water vapor pressure decreases and condensation can occur at a lower relative humidity. Condensation in some circumstances will form when the relative humidity is around 40%. As in the case with most metal surfaces, slits, micropores, and other variations on the surface can lead to a reduction in the relative humidity necessary for condensation.

This explains why you often see corrosion appearing at laps, returns, in scratches and gouges and along edges of metal surfaces. Substances on the surface, in particular hygroscopic substances such as chloride salts, will enable the development of condensation on a metal surface. Such salts tend to absorb moisture and, as temperature rises, can activate corrosion cells on the metal surface.

The rate of corrosion of metals occurs abruptly in a fairly narrow range of relative humidity. It is not a linear condition. The point where the relative humidity tends to have an effect on the corrosion rate is called the *critical humidity*. For iron, this is around 70%. This can be affected by substances on the surface of the metal and contained in the atmosphere surrounding the metal. It is good practice to store the steel in areas where the relative humidity is below this critical point.

CORROSION OF CAST IRON OR CAST STEEL

Steel and iron castings are not immune to corrosion. The same behavior applies, and protective coatings are essential. Due to the nature of the cast metal surface, different challenges can present themselves. Gray cast iron is more susceptible to corrosion than the wrought carbon steels. Relative humidity as low as 65% will corrode cast iron. The presence of chlorine or sulfur will corrode cast iron at lower relative humidity.

The graphite in cast iron is noble, that is, graphite is more cathodic in relation to the surrounding iron. In fact, on the galvanic scale, graphite is very noble. It is located at the far end of the electromotive scale even more electrically positive than gold. Refer to Table 7.3.

TABLE 7.3   Electro-potential relationship of metals in seawater.

Electrical potential of various metals in flowing seawater		
Anodic polarity	**Voltage range**	
The more active end of the Scale – Least Noble Metals	−1.06 to −1.67	Magnesium
	−1.00 to −1.07	Zinc
	−0.76 to −0.99	Aluminum alloys
	−0.58 to −0.71	Steel, iron, cast iron
	−0.35 to −0.57	S30400 Stainless steels (active)
	−0.31 to −0.42	Aluminum bronze
	−0.31 to −0.41	Copper, brass
	−0.31 to −0.34	Tin
	−0.29 to −0.37	50/50 Lead-tin solder
	−0.24 to −0.31	Nickel Silver
	−0.17 to −0.27	Lead
	−0.09 to −0.15	Silver
	−0.05 to −0.13	S30400 Stainless steels (passive)
	0.00 to −0.10	S31600 Stainless steels (passive)
	0.04 to −0.12	Titanium
	0.20 to 0.07	Platinum
The more noble end of the scale	0.20 to 0.07	Gold
	0.36 to 0.19	Graphite, carbon
Cathodic polarity		

This disparity can lead to rapid development of corrosion cells on the exposed surface of gray cast iron when an electrolyte is present on the surface. Localized corrosion cells develop where the concentrations of graphite are more cathodic in relation to the iron on the surface. Signs of this are porous, dark, rough patches of residual graphite cemented together with porous corrosion products. Figure 7.10 shows a cast iron form. The surface is spalling, and the paint is flaking from the surface. Moisture is getting through the paint coating and reacting with the steel and graphite on the surface, creating numerous corrosion cells.

Another form of corrosion faced by gray cast iron is known as graphitic corrosion, sometimes referred as *graphitization*. This corrosion condition occurs when the surface is exposed to acidic environments such as polluted urban environments. Graphitic corrosion is a form of de-alloying.

FIGURE 7.10   Cast iron corrosion.
Source: L. William Zahner.

The iron constituents are selectively leached out of the metal and leave behind a graphite rich area. Graphitic corrosion occurs just below the outer surface. This insidious form of corrosion is not readily visible on the surface. Appearance doesn't change, but you can easily scrape the surface and get the metal to flake away.

When iron or steel castings are made, the slag and oxides are removed, usually by abrasive blasting, and the surface is immediately oiled or coated with a protective paint film to prevent corrosion from occurring. A zinc-rich primer applied to the surface is recommended. The thickness of the prime coat should be relatively thick when applied to castings. Once the casting is properly prepared, the prime coat is applied directly adjacent to the metal. The dry coating thickness should be a minimum 15 μm (6 mi). For most exterior exposures where there is concern for deicing salt or high humidity, the coatings should be considerably thicker. Once dry, a compatible top coating is applied. The top coating should have ultraviolet inhibitors and should be a minimum thickness of 10–15 μm.

FIGURE 7.11   Paint on large cast fountain.
Source: L. William Zahner.

Figure 7.11 is an image of a large cast iron fountain. Nearly 100 years old, this fountain has undergone numerous repairs and restorations, some more damaging than beneficial. The old paint and corrosion were removed by abrasive blasting in the lower left image. The paint application involved a three-step process. The first coating was a zinc-rich primer for a sacrificial coating next to the cast iron. The next coating was polyamide epoxy. This coating aided against abrasion and offered a tough barrier. The final finish coating was a layer of aliphatic polyester polyurethane. This was chosen for the color, ultraviolet light resistance, and chemical resistance. In order to keep the detail of the casting, the finish thickness was adjusted so it would be thicker on the areas where moisture will be more prevalent.

COATINGS AND CORROSION

It is senseless to express the unprotected steel or iron surface and expect the surface to not develop corrosion products. The use of iron and steel, whether interior or exterior, requires a coating of some sort to prevent oxygen from coming in contact with the surface. Organic coatings, paints, and varnishes, are the most commonly used protective coatings for steel. Oils and waxes can offer temporary protection while they remain unbroken over the surface, but they wear and dry out, leaving small gaps and crack where eventually the distinctive reddish color of iron oxide will appear.

Organic coatings' primary purpose is to provide a barrier against reactants reaching the steel surface. Coatings, particularly layers, can slow the movement of ions through the layer of resin in the coating. These coatings can also hold corrosion inhibitors in proximity to the surface, such as the zinc primer coating used on the fountain in Figure 7.11. The zinc primer will act as a reservoir in the event a breach in the coating occurs. The zinc primer has zinc held in suspension until it is ready to play a role in galvanic protection of the underlying steel.

Organic Coatings

- *Act as barriers*. Slow down the diffusion of reactants to the substrate.
- *Provide ionic resistance*. Resins in the coating can slow the movement of ions.
- *Act as a reservoir*. Hold corrosion inhibitors for use in the event of a breach.
- *Hold sacrificial metals in position*. Zinc-rich coatings can provide galvanic protection.

As barriers, organic coatings can slow the diffusion of water vapor but not prevent it entirely. There is always porosity. Aging and degradation over time of the organic coating can open access to the base metal. Increasing the thickness should be a consideration where humidity or corrosive conditions are endemic. Pigments and resins dispersed throughout the organic coating not only provide color but also increase the path a reactant must take to reach the steel substrate.

Paint coatings depend on sound adherence to the surface. The thicker the coating, the better the protection against abrasion. Porosity is less. Layering compatible coatings can increase the thickness, close pores, and introduce attributes that restrict the movement of ions. The inner layer, at the steel–paint interface, can be designed for adhesion and can have added zinc to further provide cathodic protection to the steel in the event of a breach in the outer layer. The outer coating layer can provide the aesthetic of color to the steel article, as well as ultraviolet protection to the underlying coating. The layered coatings thus provide several lasting barriers protecting the steel. They are not permanent, however, and these organic coatings are subjected to the omnipresent forces of nature that will eventually win.

Adding conversion coatings on the metal surface is another important protective measure. In the form of chromates and phosphates, these conversion coatings can increase the expected lifespan of the coating. Conversion coatings produce a less-conductive region directly over the steel surface.

The chromate coatings are smooth. These must be applied over zinc-plated steel or galvanized steel. The color is an iridescent yellow–green derived from light interference as light passes through the thin layer of zinc chromate on the surface. These surfaces can then be painted or left as is. Paint bonds well to chromate conversion coatings, and these coatings achieve a good level of corrosion protection by developing an inert barrier over the steel.

The chromate coatings are heavily controlled due to the toxic aspect of hexavalent chromium. Hexavalent chromium, sometimes referred to as Cr(VI), is one of the states the chromium atom can take where the valence is a +6. This form of chromium is a strong oxidizer and will damage human respiratory and renal systems. Any chemical compound that has the chromium ion in a +6 form is considered hazardous to those who work around it and should be avoided.

There is a trivalent, Cr(III) chromate treatment that is not considered toxic and gives similar iridescent effects to the zinc-plated steel. Trivalent chromium is not as corrosion resistant, but it is not considered a hazardous compound. These coatings are amorphous, which allows post-forming operations without damaging the coating.

Zinc and magnesium phosphate coatings are rough and porous when plated over steels. This porosity can offer a tooth or anchoring point to receive the organic coating.

The zinc and magnesium conversion coatings are used extensively on firearms made of steel. These coatings are applied to the steel surface as a protective coating against corrosion. Zinc phosphate or magnesium phosphates are applied by immersing the steel parts in a hot solution for several minutes. The coatings can be applied to larger surfaces such as formed parts and plates and to galvanized steel sheet to enable paint to adhere to the outer zinc layer. These crystalline surfaces cannot undergo mechanical forming without cracking at the bends. Figure 7.12 shows the appearance of zinc and magnesium phosphate coatings.

Conversion Coatings

- Chromate and phosphate coatings on the steel surface are chemically bonded.
- Coatings provide a low reactive surface in the event of a breach in the organic coating.
- Coarseness and semiporosity benefits by adding anchoring points for the coating.

FAILURE MODES OF PAINT

For many steel-fabricated parts, the adhesion of the paint coating can deteriorate with time and exposure. The natural environment brings ultraviolet radiation, temperature changes, and the constant presence of moisture. Paints, like most everything else, are impacted by this and eventually will crack. Moisture will find its way through this crack and cause the steel to corrode. As the steel corrodes, the volume change peels back more of the paint and allows more steel to be exposed to the ever-present moisture.

Typically, the paint first fails around edges and surface changes. Figure 7.13 shows a large steel structure where the paint is failing along the edge and lower levels and moisture is getting to the steel. The piece sets in a small, planted region that receives regular watering.

Weld zones also are more prone to the initiation of paint bond failure. The plane changes create areas where the paint is thinner, due in part to the Faraday cage phenomenon, which involves tight corners or edges and the tendency of fluids to go to edges rather than the recesses.

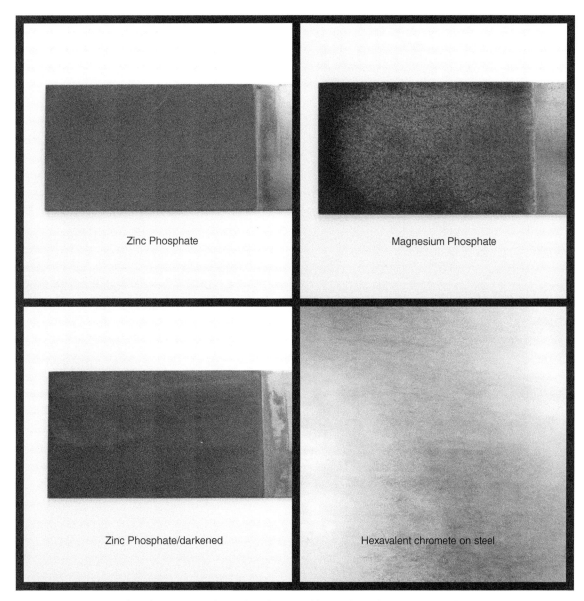

FIGURE 7.12   Zinc phosphate, magnesium phosphate, and hexavalent chromate on steel.
Source: L. William Zahner.

Figure 7.14 is a large, exterior sculpture designed by David Stromeyer. It is titled "Shim Sham Shimmy." Painting these large steel sections requires special care and control. Rust began to show where the paint was thin. The thin regions could be due to numerous factors: application discrepancies, Faraday cage effect, different drying rates on large surfaces, even abrasion from handling

FIGURE 7.13   Steel sculpture, "The Dancer." Artist Rita Blitt.
Source: Courtesy of Zahner Metal Conservation.

and setting this large, heavy form. There are methods to check the thickness of paint coatings on ferrous and nonferrous articles. Figure 7.15 shows a nondestructive instrument that measures dry film thickness. These devices use electromagnetic induction on magnetic surfaces and eddy current measurement methods on nonmagnetic surfaces to determine the thickness of coatings.

A probe is set perpendicular to the coated steel surface. The distance from the probe head and the base steel surface is a set, measured dimension. The coating thickness is determined by the difference between the electrical voltage between two coils reading the current generated by a magnetic field. One coil is in the center of the probe, and when placed on the surface being measured, the

FIGURE 7.14   Thin paint on the surface of steel sculpture.
Source: Courtesy of Zahner Metal Conservation.

magnetic flux is altered by the steel below. A second coil reads this change in magnetic inductance, and a microprocessor in the device reads this change as a thickness of a coating.

On the large steel plates, readings were taken from different zones across the surface. The areas where rust was showing through the paint were found to be thinner than the balance of the sculpture surface. The coating failed to prevent moisture from reaching the steel base.

Paint coatings are considered semiporous. There are a number of driving forces that can cause adhesion failures and the formation of rust-blistering below paint coatings. Moisture can arrive at the interface of the metal and paint coating through porosity in the coating in the form of pinholes or weakened regions along edges.

The driving forces involved with moisture progressing to the base metal under the coating are gradients in pressure created by osmotic, electrolyte migration, and thermal conditions in high humidity regions. The most common failure mode in art and architectural uses of steel coated with

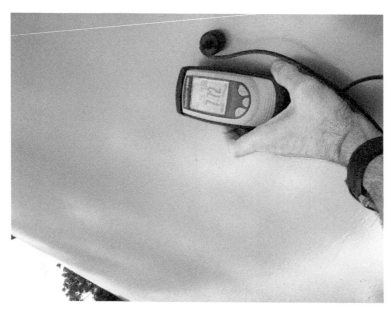

FIGURE 7.15   Paint thickness checking device.
Source: L. William Zahner.

paint are the pressures from osmotic gradients that can develop on either side of the paint film. The paint film acts as a semipermeable membrane that will allow moisture through but restrict solids. As moisture passes through the membrane, a more concentrated ionic solution forms on one side that creates an osmotic pressure that impels movement in the direction of the stronger solution.

In the case of a paint film on steel, the metal at the interface of the paint sees moisture and oxidizes. Moisture is on the external surface of the paint and may be composed of electrolytes, from humidity and fresh waters on the surface. As iron oxide concentrations climb at the paint–metal interface, a differential in concentrations on either side of the paint forms. This condition develops a driving osmotic pressure that pulls more moisture toward the steel.

If, for example, $Fe(OH)2$ begins to form on the steel and oxygen is restricted, keeping the more insoluble ferric hydroxide from forming, and there are chlorides or sulfides present, an osmotic pressure can develop and the paint will begin to blister as the oxide swells. Refer to Figure 7.16.

The blister that forms protrudes from the steel surface as the volume of the oxide is more expansive. The blister can be hard to the touch. The paint will begin to flake away as the friable oxide loses cohesion. Figure 7.17 shows a steel tube exposed to an area of high humidity. The tube was first blasted, then coated with an epoxy followed by a black urethane paint. Still, the moisture was able to reach the surface and form blisters of oxidation after several years of exposure. It is important to note that the same pipe system is used in other areas of the project where the sun would reach the surface and dry them out. These show no signs of blistering.

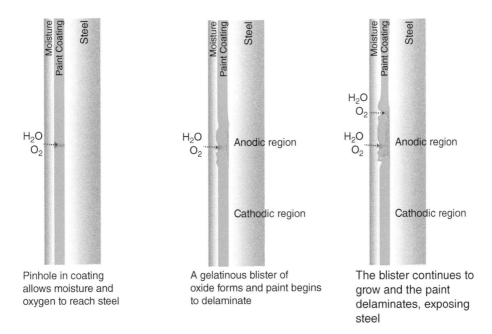

FIGURE 7.16   Blister forming under paint at steel surface interface.

Corrosion of the steel surfaces when coated with an organic coating can occur because of defects in the coatings. In moist, humid, and salt environments, the electro drive to find these defects is high. Defects can be cracks in the paint, pinholes, edges, or any number of conditions induced during the paint application, handling of the parts, and installation of the assemblies. The simple act of bolting a connection can damage the paint around the bolt. Additionally, paint coatings can change during their lifetime. Ultraviolet light degrades the coating by breaking the chemical bonds of the polymer. Also, shrinkage and differential expansion of the paint and the substrate can open cracks, allowing access to the base metal.

When chlorides are present, the corrosion vehicle is cathodic in nature as blisters or edges uplift and cations migrate under the paint at the interface of the paint to metal. The chloride ion is a negatively charged anion, attracting the positively charged cations of the metal. When this occurs, the paint will lift as the metal to polymer bonds break. As shown with the blister, osmotic pressures are generated under the paint coating. These pull in more moisture and create the swelling under the paint as the steel corrodes.

Other vehicles of corrosion also occur at the paint-to-metal interface when the conditions are right. Filiform corrosion is another form of corrosion that occurs under thin films of organic coatings on steel. They can occur on thinly plated steels, but typically they occur along the interface of the steel surface and the paint coating. They are characterized by a "worm-like" wandering track that randomly transverses under the paint coating along the steel–paint interface.

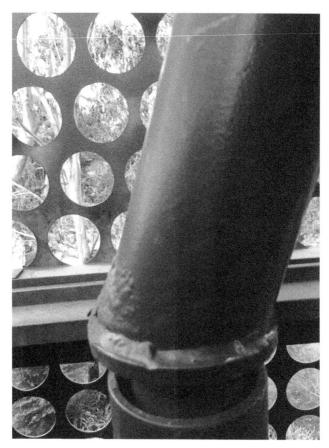

FIGURE 7.17   Blisters on pipes.
Source: L. William Zahner.

Under these small tracks you will find corrosion of the surface of the steel. The corrosion product swells and pushes and often cracks the paint coating. The track is formed as the corrosion product forms and advances as the leading edge of the track absorbs water and oxygen, continuing the process forward. Osmosis tends to remove water from the trailing edge due to low concentrations of soluble ions available as low-soluble ferric hydroxide forms in these regions. This is depicted in Figure 7.18. The paint failure at the base shown in Figure 7.13 was most likely generated from this type of corrosion.

INORGANIC PROTECTION OF STEELS

The purpose of inorganic protection of steels is to provide a barrier to diffusion of oxygen to the iron surface, provide a source of electrons and thus galvanic protection to the steel. The most common inorganic coatings used on steels are zinc and aluminum.

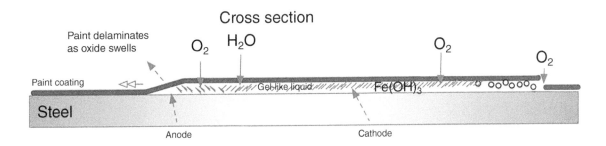

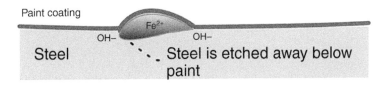

FIGURE 7.18   Filiform corrosion mechanism.

Zinc and aluminum are more anodic than steel. This means that when the two metals are coupled together, the steel will be cathodically protected by the proximity to zinc and aluminum. Zinc is the most common sacrificial metal to steels. Galvanized steel exists for this reason. The galvanizing industry is well established around the world as a preeminent method of protecting steel in all wrought forms. For that matter, the most significant use of zinc is as a sacrificial coating for steels.

Steel sheet is also often coated by aluminum and aluminum zinc alloys.

When it comes to protecting steel from corrosion, the process of galvanizing is extremely effective. Aluminum-zinc coatings as well are very effective protective measures for steel; however, these are limited to thin sheet material. Coating steel with zinc and aluminum is accomplished by dipping the steel in molten baths of the metals where it bonds metallurgically to all sides of the steel. Zinc and aluminum can also be electroplated to the surfaces.

There have been numerous studies on the protective nature and benefit of these coatings on steel. A 21-year study was performed for ASTM in 1976.[2] It compared samples of different thicknesses of hot-dipped steel in molten zinc (galvanized) along with aluminum-zinc coatings and hot-dipped, commercial pure aluminum coated steels.

The tests were performed in various atmospheric exposures. See Table 7.4.

---

[2]Townsend, H.E., and Lawson, H.H., 'Twenty-One Year Results for Metallic-Coated Steel Sheet in the ASTM 1976 Atmospheric Corrosion Tests', *Outdoor Atmospheric Corrosion*, ASTM STP 1321, E.E. Townsend, Ed., American Society for Testing and Materials International, West Conshohocken, PA, 2002.

TABLE 7.4   Test results showing the percentage of coating remaining on the surface after 21 years of exposure.

Location	G60 galvanized	G90 galvanized	Aluminum-zinc 55–45%	Aluminum – commercially pure
Coastal, low humidity Point Reyes, CA	15%	1%	0%	2%
Coastal, humid Kure Beach, NC	–	100%	12%	0%
Coastal, high humidity Panama Canal Zone	100%	100%	5%	5%
Rural, State College, PA	93%	72%	0%	0%
Urban, Newark, NJ	100%	100%	0%	0%

GALVANIZING AND METAL PROTECTIVE COATINGS FOR STEEL

Galvanizing is another common protective practice for steels, particularly thin steel sheets. Unlike plating, galvanizing makes a metallurgical bond of zinc to the steel at the surface. Steel parts and shapes are immersed in a molten bath of zinc. The longer they remain in the bath, up to a point, the thicker the coating of zinc. Figure 7.19 shows steel pipes, with welded brackets that have been

FIGURE 7.19   Galvanized steel pipes.
Source: L. William Zahner.

galvanized. Specifications for minimum thickness of zinc when galvanizing steel are dependent on the thickness of the steel. This is described in Appendix G. These forms were coated with a minimum 75 μm of zinc.

Galvanized steel coatings perform extremely well in many environmental exposure. The zinc first inhibits oxygen from reaching the steel surface. If the zinc is breached, it forms a cathodic protective barrier and sacrifices to the more noble steel alloy.

Figure 7.20 shows an art form designed by the artist Ron Fischer. The various forms were made of steel and aluminum. The steel parts were galvanized. The corrugated surface had a G90 galvanized classification, which approximates to 20 μm of zinc on each side of the sheet before it is corrugated.

FIGURE 7.20   Galvanized steel sculpture towers exposed for more than 20 years.
Source: Designed by artist Ron Fischer.

Some of the shapes were made of aluminum and some were painted as in the tall, thin steel pole. After more than 20 years exposure on a pier out in the Elliott Bay around Seattle, Washington, corrosion is appearing on some of the surfaces.

Galvanizing can be applied to thin and thick sections of steel. The benefit derived from the hot-dipping method is complete coverage of edges of the steel. Figure 7.21 shows a steel plate,

FIGURE 7.21   Hot-dipped galvanized steel plate used for tree surround.
Source: L. William Zahner.

custom cut for a tree grate cover. This is installed in a northern climate where deicing salts are prevalent. The thick zinc coating applied by the hot-dip method has performed very well.

When the galvanized coating is too thin, it will begin to fail as the zinc layer is consumed from corrosive attack. Figure 7.22 shows a thin steel corrugated sheet that had been galvanized. The thinner the sheet, the more economical the material is, but also the thinner the zinc coating is. Thin steel sheet has a limit on how thick the zinc coating can be. These sheets show signs of failure most likely initiated in the storage process. The pattern of corrosion indicates they may have gotten moisture between the corrugated sheets while they were stored. Once installed, they began to corrode more rapidly.

When moisture gets between galvanized steel sheets and is allowed to remain, it begins a deterioration process. Indications of this are a whitish stain that forms on the galvanized steel. This coarse white stain is zinc hydroxide. Refer to Figure 7.23. It is not easily removed. This zinc hydroxide is insoluble and actually protects the metal to a degree. When it comes off or doesn't completely form, it weakens the zinc layer and corrosion of the underlying base metal will occur.

Other coatings that are applied in the molten state to the steel surfaces include aluminum and zinc-aluminum alloys, zinc, and magnesium and tin coatings. The zinc-aluminum alloys are another common coating on thin sheet steel products. These are alloys composed of 95% zinc

FIGURE 7.22   Corroding corrugated galvanized steel sheets.
Source: L. William Zahner.

FIGURE 7.23   White zinc hydroxide forming on galvanized steel sheet.
Source: L. William Zahner.

and 5% aluminum. Similar to galvanized coatings, these are applied to both surfaces by hot dipping the steel in a molten bath.

Tin once was common on thin steel sheets and is still used in the can industry. Tin plates were a common roofing material used on homes and cornices of buildings during the late 1800s and early 1900s. The steel plates were dipped in molten tin to produce a thin coating over the surface. Refer to Figure 7.24. Tin sealed the steel surface from moisture and air but did not offer galvanic protection to the base metal. In fact, exposed steel would sacrifice to the tin in the presence of moisture. Edge exposures would often rust. These steel plates were thin, and therefore, once corrosion was initiated, the steel would virtually dissolve. Many of these tin-plated roofs were coated with lead paints for added protection. The lead paints were red in color. Restoring them involves careful replacement, usually with more modern metals and metal coatings because of the lack of availability of tin-plated sheet and the elimination of lead products in building construction.

PLATING

Other methods such as plating, if performed correctly, will protect the ferrous surfaces from corrosion. Steel automobile bumpers are plated with nickel and chrome to establish protective barriers over steel. Many steel appliances are also chrome, nickel, or even copper plated, in order to give the beauty of one metal on the strength of steel. Plating involves a pickling process to remove the oxide from the surface of the steel part and immersion in an ionic bath of a metal salt. The steel part is

FIGURE 7.24   Restored tin-coated steel roof on the Ottawa, Kansas, train depot.
Source: L. William Zahner.

made the anode and metal ions in the solution plate onto the exposed steel surface. Zinc, nickel, copper, chrome, and even silver can be easily plated onto the surface of steels.

WEATHERING STEELS – HOW THEY DIFFER FROM OTHER STEELS

The weathering steels are remarkable alloys of steel from the perspective of corrosion. Weathering steels form a complex layer of oxides and hydroxides over the entire exposed surface. The complex layer will protect the underlying steel from further corrosion as it develops into a thick, electrochemically inert oxide. Figure 7.25 show a weathering steel surface that has developed over 40 years. The color is a deep brown with a slight purplish hue. This surface will not rub off, flake, or stain adjacent surfaces. It is a very durable and hard steel surface.

Once the oxide has fully developed it is very chemically stable. Moisture has no effect and even chloride deposits are resisted by the oxide layer. However, when chloride deposits are allowed to stay on the surface, in the case of horizontal surfaces, the surface will react and can form iron carbonates that will sluff off the surface.

FIGURE 7.25   Forty-year-old weathering steel surface on the Zahner Kansas City Factory.
Source: L. William Zahner.

Weathering steels are a family of steel alloys that contain special low concentrations of various elements that impart a specific behavior to the oxide making it more adherent and stable. One of the elements that aid in the development of the resilient oxide layer is copper; thus, the name *copper-bearing steels* is sometimes used to describe these steels. Also, small amounts of chromium, nickel, phosphorous, and manganese are present. The weathering steels are harder and stronger than most other steels, possessing a yield strength substantially greater than many low-carbon steels. Bridge construction in the United States uses weathering steel beams on over 40% of the new bridges under construction. These are high-performance grades of weathering steel and their success is due mostly to the low maintenance benefits they offer. Diesel fumes and sulfur dioxide from acid rain or combustion products will not have a detrimental effect on weathering steels.

The use of weathering steels is not immune to all corrosion concerns. The oxide layer grows as the surface undergoes periods of wet and dry cycles. The key to success is alternating periods of wetness followed by drying out so the oxide layer forms sufficiently to create the desired corrosion resistance. When the surface remains wet, the initial development of ferrous hydroxide forms away from the surface and does not bond. Submersing the weathering steel in water or even moist soils will develop the oxide, but it will form away from the surface. Saltwater and deicing salts allowed to reach the surface before the thick oxide develops will cause a weak oxide to form. The characteristic color of this weak oxide is a yellowish-orange color, corresponding to the soluble ferrous oxides. These will break away from the surface and stain adjacent substances.

If the surface oxide dries out too rapidly, a weakly bonded oxide will also form. This can lift from the surface and flake off. It is the right amount of moisture, time of wetness, and temperature that are needed to develop the thick, adherent oxide layer.

WEATHERING STEEL PERFORMANCE IN VARIOUS ENVIRONMENTS

The performance of weathering steels in different environments has been examined by organizations around the world. In the 1970s and 1980s an extensive test was performed on ASTM A588 weathering steel.[3] This is the plate form similar to the K11430 and K12043 alloy described in Chapter 2.

Samples of the steel were exposed for eight years in four locations around North America to correlate to specific environments. Kure Beach, North Carolina, would be one site to represent the humid marine environment; Saylorsburg, Pennsylvania, to represent a typical rural environment; Bethlehem, Pennsylvania, to represent an industrial environment; and Newark, New Jersey, would represent a typical urban environment.

The tests show that the majority of the corrosion on the surface happened in the first two years and after that, the weathering steel experienced a rapid decrease in the corrosion rate. This was the case at all sites. The samples were examined at different points. After the eight years of exposure the results are as follows (Table 7.5).

This is quite remarkable when you realize that most of this happened in the first two years. The study goes on to postulate further corrosion of the surface will be so minimal that in the most corrosive environment evaluated, Kure Beach, North Carolina, it would take nearly 40 years before the weathering steel would lose 250 μm (0.1 in.).

---

[3]Townsend, H.E., Zoccola, J.C., 'Eight-Year Atmospheric Corrosion Performance of Weathering Steel in Industrial, Rural, and Marine Environments', *Atmospheric Corrosion of Metals*. ASTM STP 767, S.W. Dean, Jr. and E.C. Rhea, Eds., American Society for Testing and Materials, 1982, pp. 45–59.

TABLE 7.5   Recorded average loss in thickness of weathering steel in various exposures.

Environment	Location	Average loss in thickness (μm)
Marine	Kure Beach, North Carolina	100
Rural	Saylorsburg, Pennsylvania	70
Industrial	Bethlehem, Pennsylvania	70
Urban	Newark, New Jersey	40

These findings correlate well with what is being experienced in other areas. Another study performed on a preweathering steel façade, less than 20 m from the shore in British Columbia found similar results.[4] The project was clad in nominal 2 mm thick preweathered steel meeting ASTM 606-4. A control sample of the original material as well as interior panels, not exposed to the exterior environment were used for comparison. The ocean facing façade is south facing, so it receives both the drying conditions offered by the sun and humid, salt-laden air of the ocean. The project has been up for close to seven years. It was concluded that the oxide actually grew, making the panels slightly thicker by as much as 0.1 mm. This would correspond to the expansive nature of the oxide as compared to the base metal. Refer to Figure 7.26.

The oxide was removed to the base steel on both a control sample and an area of the exposed sample on the roof. The measurement shows at best, the base steel reduced in thickness by approximately 0.05 mm.

Another extensive study on weathering steel tubular forms used for tower construction was perform and published in 2002.[5] There were 23 sites examined over a 15- to 30-year period. This study concluded that the natural oxide development is less in cold-dry regions than in moist, warm regions. Examination of weathering steel forms in Florida exposure found the oxide growth was most rapid in the first two years then gradually slows and eventually ceases. The study suggests that after 22 years the thickness of the oxide grows to approximately 0.30 mm with the majority of this thickening occurring in the first two years.

The volume of each of the common forms of oxides expands to approximately two times that of the base metal. Therefore, you would expect a thickening of the metal as the surface oxide develops and consumes, as it were, some of the iron on the surface.

Temperature and humidity play a major role in the both the time of developing the oxide and the soundness of the oxide. When the surface is too warm, the oxide dries out and lifts from the surface. This is because the initial oxide forms off the base metal. As it grows with moisture present, iron diffuses outward and forms more adherent oxide that is stabilized from the copper and nickel in the

---

[4]This study was undertaken by the A. Zahner Company performing an evaluation of its Solanum™ preweathered steel. The use of an ultrasonic probe and measurements comparing control samples to actual by micrometer over numerous locations on the façade arrived at the findings. 2019.

[5]Hoitomt, M.L., 'Performance of Weathering Steel Tubular Structures', *Outdoor Atmospheric Corrosion*, ASTM STP 1421, H.E. Townsend, Ed., American Society for Testing and Materials International, West Conshohocken, PA, 2002.

FIGURE 7.26   Private residence on coast of British Columbia, Canada. Designed by Tony Robins.
Source: by Ema Peter.

alloy. See Figure 7.27. Here, the surface oxide developed quickly but has lifted from the surface and is flaking off, exposing base metal underneath.

If the relative humidity is low, the oxide also dries out rapidly and iron does not appear to diffuse from the base metal into the oxide effectively.

When the oxide forms properly on weathering steels, the resultant loss of base material is negligible even under the most corrosive conditions. The reduction of structural integrity due to corrosion is manageable. That being said, however, instances where thin weathering steel was used as wall panels, have failed. The failure was due to conditions that allowed both surfaces to be wet.

Early uses of weathering steel as thin siding made by roll forming the thin sheet into various panel profiles proliferated in the late 1960s and early 1970s. The desire was to use thinner steel coil material to arrive at an economical, attractive surface material. Thinner sheet made it possible for roll forming light-gauge metal panel systems, and this accommodated the speed of production and thus reduced cost opportunities. Some weathering steel panels were manufactured as thin as 0.8 mm (0.032 in.). Unfortunately, they were used as siding panels in climates that experience rains and humidity. Both surfaces of the steel experienced oxidation, and after a number of years the

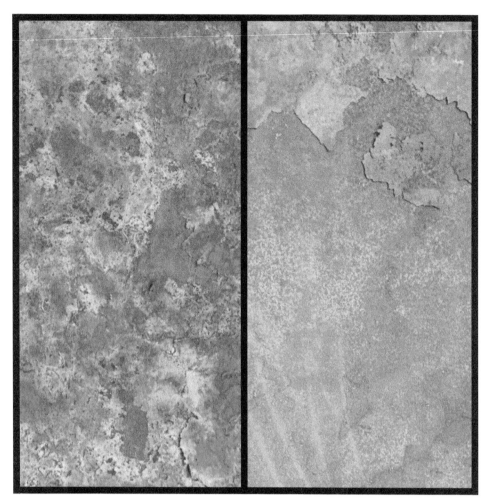

FIGURE 7.27   Lifting and separation of the thin oxide layer from the base metal.
Source: L. William Zahner.

thinning of the metal was sufficient to affect the structural integrity. The oxide is porous and does not provide any structural capacity without sufficient base material to support it. These surfaces failed prematurely, and for a period of time, the weathering steel usage as an architectural cladding material was curtailed.

It is recommended that weathering steel alloys be minimum 2 mm.
    It is recommended for thin surfaces of weathering steel that the reverse side be painted with a zinc-rich primer to prevent corrosion.

From a corrosion perspective, weathering steel is a superior choice for all typical and many severe environments. It is not omnipotent, but once the thick surface oxide develops, few conditions will deteriorate this metal. Most damage develops from improper detailing or misguided expectations.

When developing systems using weathering steel, detail the surfaces to drain. Do not have areas where water can pool and stand. Even on the preweathered surfaces, if water is allowed to pool it will discolor the steel and potentially deteriorate the steel. As water rests on the surface, iron ions will diffuse into the water. Signs of this are a brighter yellow color on the darker reddish brown background. The yellow color is due to the formation of iron (III) hydroxide. As the water dries, this is redeposited on the surface, but each time the water pools, more will form. Figure 7.28 shows a preweathered steel surface where the top surface was sloped to create the artistic form. Unfortunately, this directed water to this area and created the orange contrasting streaking.

FIGURE 7.28   Streaking from drainage directed over the surface.
Source: L. William Zahner.

FIGURE 7.29   Sculpture made from weathering steel. Ewerdt Hilgemann, artist.
Source: Ewerdt Hilgemann artist.

Storing weathering steel where water will collect will also generate this condition as iron ions diffuse out and combine to form the yellow/orange iron (II,III) oxide. Figure 7.29 shows a beautiful sculpture made from weathering steel. The artist, Ewerdt Hilgemann, imploded this large weathering steel form. The form is over 7 m in length, and handling required it to be set on its side. Moisture collected in the imploded shape, creating these interesting evaporation lines. These will eventually darken as the oxide thickens, but the contrast will be apparent for many years.

Avoid saltwater spray wetting the surface constantly. This will also leave the yellowish streaks and the oxide will deteriorate. Same with deicing salts. Deicing salts will cause a spotty and streaky stain to develop. This stain is a brighter color and is a sign the thickened oxide is not forming. Once the oxide develops, however, it is more resistant to the corrosive effects of saltwater, as shown in

FIGURE 7.30 "The Comb of the Sea." Artist Eduardo Chillida. More than 40 years exposure along the western end of the La Concha Bay, San Sabastian, Spain.
Source: Artist Eduardo Chillida.

Figure 7.30. These rugged weathering steel forms, called, "The Comb of the Sea," designed by the artist Eduardo Chillida, have been standing along the rocky coast of La Concha Bay in San Sebastian, Spain for over 40 years. They are made of thick plates of weathering steel welded and formed.

When Working with Weathering Steels Avoid the Following

- Pooling water on the surface
- Saltwater spray constantly striking the surface
- Deicing salts
- Submersion in water
- Submersion in moist soil
- Bird and other organic matter
- Wood products directly adjacent
- Copper draining onto the steel surface

The misguided expectations of weathering steels occur when the metal is expected to perform like a paint. Often when streaks develop on the exposed surface the assumption is the weathering oxide is failing. Streaking will develop around edges, perforations, or regions where moisture is collected and directed. This streaking is always of a contrasting, lighter color. Edges and perforations collect moisture from adjacent regions.

In the case of preweathered steel, the edges of perforations and cuts are difficult to accelerate weathering. The oxide does not develop as consistently. This creates regions where the initial oxidation still must form. This initial oxidation is the ferrous, iron (II, III) oxide and hydroxide, which is less adherent and soluble in water. It has to form and stabilize, and until this occurs, it will streak. Figure 7.31 shows a waterjet cut edge on preweathered decorative steel panels. The

FIGURE 7.31   "Crown of Thorns" fence showing streaks from cutout designs.
Source: L. William Zahner.

horizontal streaking is due to the panels having been stored outdoors at a horizontal bias. The vertical streaking at the cuts is from the base steel at the cut. Moisture has collected, reacted to form corrosion products, and is weeping out. This will occur until the more insoluble oxide form develops. Depending on the climate, you should expect this to continue for as much as a year.

You can speed up the process by regularly wetting the surface with clean water and allowing it to dry. The oxide grows and changes as the surface and edges are wetted and allowed to dry out in cycles. Figure 7.32 shows this same "Crown of Thorns" fence after a few years of exposure. This fence was not preweathered but allowed to weather naturally.

Preweathering in a controlled environment will help eliminate much of the streaking. You should expect some along the cut edges and, if perforated, bleeding from the edges of the perforated holes. On thick plates of weathering steels, there is more area around the perforated edge that must oxidize. These areas take a bit longer, and while the oxide develops streaking will occur. Figure 7.33 shows a 4 mm thick plate with streaking on the surface as the perforations develop the more stable oxide.

Eventually, this streaking will blend in and the contrast will be less (Figure 7.34). This may take a year or so. It can be accelerated by cycling wet and dry conditions over the surface. A few times a day over several days when it is warm will usually reduce the contrast.

FIGURE 7.32   "Crown of Thorns" fence after a few years' exposure.
Source: L. William Zahner.

FIGURE 7.33   Streaking on preweathered surface with perforations.
Source: L. William Zahner.

ENVIRONMENTAL EXPOSURES AND THEIR EFFECT

Any discussion involving the corrosion of mild steels in normal atmospheric exposures centers on the efficacy of the coating used to protect them. Simply put, mild steels, whether cast or wrought, regardless of form, will develop corrosion on the surface if left unprotected. Still, it is always impressive to see steel that has been exposed for decades, even centuries, as the Iron Pillar of Delhi has demonstrated or as shown in Figure 7.35 rural steel fencing that has survived for years as a tree grew around and engulfed it. Still showing strength and durability.

CARBON STEELS

Carbon steels, whether structural or aesthetic, today are always coated either with an inorganic material such as zinc or porcelain or with an organic paint coating. How these coatings perform in the environment, and how the metal-to-coating interface is produced play into determining how the overall surface will withstand and perform as designed in various environments. There are entire industries surrounding the efficacy of various coating systems. The material, in this case steel, is only one of many substrates that can be used.

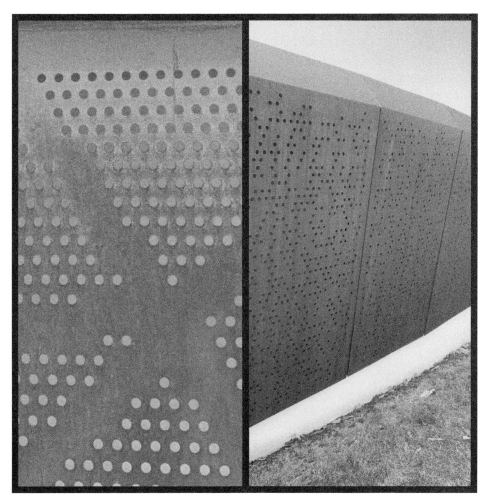

FIGURE 7.34   Streaking contrast is less apparent or nonexistent after a few months of exposure. *Source:* L. William Zahner.

There is a myriad of paint coatings used on steels. Similarly, there are a vast number of primer and preparations needed for these coatings. There are basic requirements to be followed in all the coatings for the carbon steels, organic, and inorganic. Those are:

- Clean, dry surface. Free of scale, oxides, grease, oils, and moisture
- Mutually compatible primers and paints
- Prepared surface, including edges, welds, returns, and piercings

FIGURE 7.35   Iron Pillar of Delhi and barbed wire fencing engulfed by a tree.
Source: By Shutterstock.

Paint failures typically fail around edges and welds. These areas often get the least preparation. If the weld is performed in the field or after initial abrasive blast, consider wire brushing to remove slag and roughen the surface. If the primer coating and the paint coating are not compatible, failure is assured. It is important to establish a relationship with the paint coating supplier to ensure a clear understanding of what it is to be painted and in what environment the paint is expected to perform (Table 7.6).

TABLE 7.6   Coatings used on steel.

Organic coatings

Linseed oil-based paint	Poor chemical resistance Poor weatherability	Good adhesion
Oleo resinous varnishes	Poor chemical resistance Poor weatherability	Good adhesion Fast drying
Alkyds	Poor ultraviolet Difficult to touch up damage	Good adhesion, thick coatings
Modified alkyds	Poor ultraviolet	Good durability and color Good adhesion
Phenolic lacquer	Good chemical resistance	Yellow out as they decay
Epoxy	Good chemical resistance and good flexibility	Yellow out in certain light exposures
Urethanes	Requires primer to adhere to metal; UV resistance is fair	Urethanes are flexible while the polyurethanes are harder
Vinyl coatings	Good flexibility and adhesion	Can crack and yellow
Acrylics	Hard, sometimes brittle coatings with good UV resistance	Color choices are many and application methods are both wet and dry
Polyesters	Good flexibility and good ultraviolet light resistance Poor alkali resistance	Polyester powder coatings offer a large variety of colors
PVDF	Premier among coatings Excellent UV and chemical resistance	Lack flexibility
Bituminous	Chemical resistant coatings Thick	Generally concealed from view Black, tarry look
Zinc-rich	Excellent base coat for structural forms	These are thick coatings only available in gray tone
Silicone	Chemical and heat resistance Can be blended with other coatings types	Difficult to remove
Inorganic coatings		
Porcelain	Excellent chemical resistance Excellent chemical resistance	Inflexible Edges and the back side can be a problem with corrosion
Galvanized	Excellent weathering, good chemical resistance	Doesn't take paint well without preparation
Aluminum–zinc coatings	Superior weathering Good chemical resistance	Doesn't receive paint well
Plating	Very durable Metal plating has good coverage	Can be porous and allow moisture access to base metal

Carbon steels, whether used inside or outside, require a coating to prevent oxidation and corrosion of the surface. The next group of steels, the high-strength, low-alloy HSLA steels commonly known as the weathering steels, painting is not required, nor is it desired in order to achieve a deep, rich brownish-red color and corrosion-resistant surface.

For clear coatings used on steels, they require good adhesion without the need for a primer. The surface must be as clean as possible. For the cold rolled steels, the surface is smoother than hot rolled. These can pose some adhesion issues. If you can add some tooth to the surface, successful application of the clear coat will improve. Clear coatings on the bare steel are strictly for interior applications.

For exterior applications the most common and cost-effective coating processes on steel is the coil coating of sheet material with high-quality paints. Coil coated steel is extensively used across industries because it combines the strength, ductility, and economy of thin steel, with the flexibility, barrier protection and aesthetic functions of paint coatings. Coil coated steels are still subject to the same electrochemical conditions that lead to corrosion in our built environment, but they employ several additional mechanisms to protect the underlying steel and extend the useful life of the steel form:

- A conversion treatment to the steel surface prior to application of the paint coating
- The layering of galvanized, aluminum, or other "self-healing" barriers directly on the steel surface
- Adding top layers of clear or pigmented coatings

These additional layers are applied continuously on a strip of steel sheet, which introduces economy and consistency into the system. From a corrosion protection context, these additional mechanisms effectively protect the steel from the drive of electrochemical forces present in our natural surroundings. The isolation and protection of the steel from the transfer of ions, exchange of electrons from cathodic to anodic regions on the surface, the migration and diffusion of particles across the barriers is effectively controlled by these mechanisms.

Keep in mind, however, these coils of metal will undergo shearing into sheet, forming, stamping, piercing, and cutting operations as they are shaped into the useful articles of commerce. These processes will expose edges, compress, and stretch layers, even open microscopic cracks allowing pathways to the steel substrate to eventually form. The "self-healing" layers, typically derived from sacrificial metal layering, will provide some edge protection and resist the electrochemical forces until they are spent.

There are numerous siding panels and insulating metal sandwich panels manufacture from coil-coated steel sheet that have performed effectively for 50 years in urban and marine environments. They date back to the late 1960s and early 1970s. During this period, coil-coating systems on steel were being perfected as paint coatings of exceptional durability were formulated. The fluoropolymer coating known as polyvinylidene difluoride (PVFD) came into production. This is an outstanding coating for metals and is considered the top architectural coating in the industry. Originally designed for coil coating applications, it now is spray applied and can be both powder coated and wet coated.

Still, with the advent of the high-quality, durable coatings, the most important aspect of any paint coating on steel is the interface between the steel and the paint. How the steel is prepared, what form of pretreatment, conversion coating and primer are as important or more so than the paint coating applied.

Some of the best-performing systems involve several process layers on a steel sheet. The product first can be galvanized or galfan coated. Galvanized steel is the coating with a thin layer of zinc. Another method may involve galfan, the name given to coating by hot dipping similar to galvanized but into a bath of an alloy of zinc and aluminum. Both of these metal coatings are applied to both sides of the steel sheet. These coatings produce a barrier at the steel interface as well as sacrificial protection to the base metal. These coatings, mainly composed of zinc, cathodically protect the steel in the event of a breach in the paint coating.

These metal coatings have to be prepared to receive paint and allow the paint to bond to the surface. Oil-based paints can create conditions where the zinc coating oxidizes to form carboxylates when exposed to weathering conditions. The oxide of zinc reacts with the fatty acids in the oil-based paints and can form a saponification condition, essentially the paint lifts off the zinc. Once the metalized coatings are prepared, the primer is applied.

Prime coatings are typically from 7 to 20 μm in thickness. They can be based on urethanes, acrylics, epoxy, and polyester systems. It is very important that these primers are fully compatible with the top coating system used.

Top coating can be single or multiple systems composed of polyvinylidene fluorides, polyesters, urethanes, and plastisol type coatings usually around 20 μm in thickness. These coatings contain pigments for esthetic purposes as well as ultraviolet inhibitors and other additives for chemical resistance.

STEEL USED IN DIFFERENT ENVIRONMENTS

You want weathering steels to corrode. Weathering steels are designed to oxidize to a point until the oxide that develops on the surface becomes unreactive with the atmospheric conditions that surround them. In exposures around the globe where the metal is used as a cladding or roofing surface, or a sculptural form, different conditions will be experienced by this surface oxide. It will not behave the same in a rural exposure as it will on a seacoast exposure or industrial urban environment.

The international Standards Organization, ISO, has defined five distinctive environmental exposure categories. These categories help to identify if a steel, coated or weathering, is suitable for the exposure and what measure may be necessary to keep the surface performing adequately. The categories are as follows (Table 7.7):

These classifications help to provide a direction for the coating system to be employed and an approach to design and maintenance. Most paint systems in use today can be suitable for C1 and C2 conditions. Galvanizing, the common metal on metal coating system, is suitable for these conditions as well. Preweathered steel that has been burnished during the process of weathering to remove all loose particles and where the surface oxide is fully developed, is suitable for interior exposures. Figure 7.36 shows an interior application where the preweathered steel is used. The loose particles have been removed.

TABLE 7.7   ISO environment classifications.

Category	Environmental descriptions
C1	Indoor environment. Arid desert environment.
C2	Indoor environment, unheated. Rural environment.
C3	Indoor environment, high humidity. Northern coastal. Clean urban.
C4	Indoor environment, chemical exposure. Polluted coastal. Urban north.
C5	Industrial. High humidity, high pollution. High salt-exposure environments.

FIGURE 7.36   Interior partition made of 3 mm weathering steel. Detail of intersection.
Source: L. William Zahner.

Organic pigments have a limited life expectancy due to the degradation of the chemical bonds that make up the polymer. The preweathered steel, however, has a mineral surface that will not degrade from ultraviolet radiation, moisture, or chemical attack.

Moving to C3 conditions, for painted steels the preparation and the primer used on the steel before the finish paint is important, as is the detailing of connections and joints. Ponding water on the surface will degrade the paint as it seeks out pores and cracks in the coating system. Preparing the surface for paint by following the paint manufacturers suggestions, proper surface preparation such as abrasive blasting, and degreasing prior to coating. Primer and face coating should be compatible and of adequate thickness to protect the underlying metal.

Weathering steel used outdoors and preweathered steels indoors and outdoors will perform very well with little maintenance requirements. Deicing salts and sea spray allowed to remain on the surface can cause the surface of the weathering steel to decay and show the lighter color oxide on the darker background.

As you move into the C4 conditions and C5 conditions a design may want to consider paint coatings that provide chemical resistance. In these exposures, the preparation of the steel surface, the appropriate primer and top coating are very important. Adding some form of cathodic protection may be worth considering. Many museums consider cathodic protection along with epoxy primers or zinc-rich primers for sculpture works in contact with soils. Cathodic protection involves connecting the sculpture to an electrical circuit where the sculpture is the cathode and is electrically connected to an anode set remotely. A rectifier inducing a flow of electricity may be incorporated. In this way, all the corrosion is centered at the anode as it protects the sculpture.

In all environments some cleaning regime should be considered. Indoor applications simply wiping the surface occasionally with a clean rag and mild detergent or deionized water.

Some exterior environments, on the other hand, can make the process of regular cleaning very difficult and expensive. In these instances, it may be worth considering a durable surface. Weathering steel in both the natural weathering surface and the preweathering surface requires little maintenance, just the occasional pressure washing of the surface to remove debris and bird waste. Steam is also a good cleaning method to get deep into the porous surface. Paint and adhesives can be removed with the use of organic solvents without damaging the oxide surface.

TYPES OF CORROSION

Dr. Herbert H. Uhlig, the renowned chemist and corrosion expert, defined corrosion as the destruction of a metal by chemical or electrochemical reaction with its environment. We are familiar with rusting steel and the crumbly residue as it corrodes. We are familiar with the patina on copper as it changes colors and forms a thick coating of oxide particles. These are the changes Uhlig refers to as chemical and electrochemical reactions as metals interact with the environment and combine with the atmosphere.

Corrosion is the tendency of a metal to return back to its mineral state, the state it was originally in before refinement. Corrosion is, in reality, a thermodynamic process that cannot be stopped, only slowed.

TABLE 7.8   Corrosion types used in art and architecture.

Corrosion type	Faced by art and architecture
Uniform	Common
Galvanic corrosion	Common
Pitting	Common
Intergranular	Common
Weld corrosion	Common
Crevice corrosion	Common
Line corrosion	Rare
Corrosion fatigue	Rare
High temperature	Rare
Stress corrosion cracking	Rare
Fretting corrosion	Rare

For art and architecture, the environment is more benevolent, and many uses of steel, other than the weathering steels, consists of applied organic or inorganic coatings to interface with the environment.

There are several types of corrosion faced by steel. Of these, only a few are commonly faced by artistic and architectural uses of the metal. Granted, all of the types listed are encountered by steel or at least should be considered as the metal is designed into a project. The realities are, however, some of these will never be confronted in art and architectural uses of the metal steel (Table 7.8).

For corrosion to occur on a metal, there must be an electrochemical reaction on the surface. The electrochemical reaction that causes corrosion begins as soon as a cell circuit is formed. A cell circuit is a movement of electrons from a negatively charged point, the cathode, to a positively charged point, the anode. This can occur in the case of dissimilar metals, where one metal has a different electro-potential than another. In this case, the more noble metals are more electrically positive. When coupled with metals that are more negatively charged, reference electromotive chart Table 7.3, the more negative will corrode in relation to the more positive or noble metal. It also can occur on the surface of the same metal. Metals can establish localized differences in electrical charge on their surfaces when they are exposed to certain atmospheric conditions that possess the ability to establish an electrochemical reaction. Depending on the environment, the electro-potential differences can be very small and thus slowly affect corrosion behavior, or they can be significant in the case of pollution or salt exposure. The vehicle of corrosion can also be autogenic in nature. Once it starts, it will continue until the circuit is broken.

UNIFORM CORROSION

Uniform corrosion happens over the surface or a large portion of the surface in a uniform attack of the steel. From an industrial context, uniform, or general corrosion relates to the attack of the entire surface of a metal when exposed to certain chemical agents. In the case of steel, water resting the surface can cause uniform corrosion. Figure 7.37 shows carbon steel plates stored outdoors and uniform attack on the surface.

Steels, before they are coated with zinc or paint often are protected with a light mill oil. This mill oil will lend protection against uniform corrosion for a short period of time if the steel is stored in an indoor dry, low-humidity place. Unless this is allowed to continue for several months or exposure to chlorides or pollution occurs, the rust is superficial. For example, when the surface is intended to be hot dip galvanized, the metal form with the light corrosion will be dipped in a pickling bath of acid. This will remove the oxides from the surface and make the surface receptive to the molten zinc where the zinc and iron will metallurgically bond.

If the steel fabrication is to be painted, the uniform corrosion is removed by abrasive blasting the surface. The abrasive blast creates surface roughness, and this gives anchor points for the paint to key into the surface.

FIGURE 7.37   Uniform corrosion on steel plates exposed out of doors.
Source: L. William Zahner.

FIGURE 7.38 Weathering steel first developing the uniform corrosion over the surface.
Source: L. William Zahner.

Uniform corrosion will not have a detrimental effect when it occurs on weathering steels. It is uniform corrosion that first develops over the surface of weathering steels and this initiates the process of formation of the protective oxide layer. Figure 7.38 shows a weathering steel surface with a form of uniform corrosion developing over the surface.

GALVANIC CORROSION AND GALVANIC PROTECTION OF STEELS

On exterior applications in art and architecture, we tend to focus on dissimilar metals and the potential effect of galvanic corrosion. Steel is anodic when coupled with metals such as copper, brasses, and stainless steels but not so with zinc and aluminum. Zinc and aluminum will be sacrificial to steels and provide cathodic protection to the steel.

Steels, both low carbon and the weathering steels, in contact with copper and brass, will undergo rapid corrosion. Steels, similar to all metals, will undergo galvanic corrosion when coupled with any metal more positive in the electromotive or galvanic series. All metals, when immersed in an electrolyte possess an electric potential. This is the energy that is released as corrosion is initiated. Depending on the liquid environment, when two different metals are present, they generate a current that passes through the electrolyte. It is the direction of this current that determines which metal will undergo corrosion. This is galvanic corrosion.

Table 7.3 shows a common relationship of various metals exposed in a salt environment. For the exposure in most architectural environments this is the arrangement of metals, however the voltage potential may change depending on a number of factors. Temperature, concentration of chlorides, and aeration in the electrolyte all will have an effect on what the actual voltage measure will be. The chart shown in Table 7.3 is more of a directional indicator.

For example, the reason galvanized steel offers such good corrosion resistance is because the zinc is sacrificial, more electrically negative than the steel. Copper and brass are more positive than steel and thus they corrode steel when in combination in an electrolyte. Galvanic corrosion is most dramatic at the region where the steel touches the more cathodic metal.

Figure 7.39 shows a copper handrail attached to the steel support. The steel has undergone extensive galvanic corrosion because of the coupling of the two dissimilar metals. The copper, on the other hand, is in good shape.

FIGURE 7.39   Corrosion of steel supporting a copper handrail.
Source: L. William Zahner.

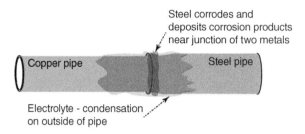

Steel corrodes and
deposits corrosion products
near junction of two metals

Copper pipe

Steel pipe

Electrolyte - condensation
on outside of pipe

FIGURE 7.40   Galvanic corrosion at joint between copper pipe and steel pipe.

Copper and copper alloys will create corrosion cells near the area of contact when moisture is present. Many a copper pipe connected to steel piping is a testament to the corrosive effects of this galvanic couple (see Figure 7.40). It is recommended to avoid contact between steels and copper alloys.

When a galvanic couple is developed, the anodic metal is said to be sacrificial to the cathodic metal and often prevents the cathode from corroding. So, in the instance, as depicted in Figure 7.40, if a steel pipe is joined to a copper pipe and an electrolyte is present, the steel will corrode and deposit iron oxide around the joint. Little to no corrosion will occur to the copper pipe while the steel pipe will undergo significant and rapid corrosion as long as the electrolyte is present, and electrons are allowed to flow through the electrolyte. Note, condensation is the vehicle of ion flow. If the pipe is dry, the rate of corrosion would be very slow, if at all.

There are two processes involved in galvanic corrosion: one is the corroding or dissolution of the anodic metal. In the case above, this is the iron pipe, and the reaction is

$$Fe \rightarrow Fe^{2+} + 2e$$

Here the iron ions are dissolved into the electrolyte.

The second process is that the dissolved oxygen found in the electrolyte goes through a process referred to as reduction. The oxygen, along with the water molecule and these electrons that are now released from the iron, react as follows.

$$O_2 + 2H_2O + 4e \rightarrow 4OH^{4-}$$

When this happens, the iron goes into the solution as a positive ion and combines with the negative hydroxide and forms iron hydroxide or rust. This comes out of solution and deposits on the pipe.

This is referred to as an oxidation-reduction reaction. This is the basis around galvanic corrosion. It is an electrochemical action. There is a flow of electricity through an electrolyte by means of ions in the solution. An electrochemical reaction is defined by the transfer of electrons in a chemical reaction known as oxidation and reduction. Stripping electrons from the iron is called oxidation. It is not necessarily associated with oxygen.

Adding electrons, in the second equation, is known as reduction or cathodic reaction.

The anodic metal by itself may not react in a similar electrolyte when not coupled with the cathodic metal, but when coupled, the corrosion process will continue as long as the electrolyte is present.

Additionally, the two metals do not necessarily need to be physically connected but only need to be electrically connected. The most common galvanic or bimetal corrosion condition occurs when they are physically connected; however, if the two metals are joined by conductive media, galvanic corrosion can still occur if ions form at the anode, as with the iron pipe. Here oxidation is occurring, and the iron is corroding.

At the same time, there must be an acceptance of the electrons at the cathodic site, the copper pipe. The corrosion occurs only at the site where the two metals connect in this case. The entire steel pipe is not affected – only the area near the copper where the ion flow is the greatest.

The strength or magnitude of galvanic corrosion is dependent on several conditions. The electromotive scale shows an electric charge potential. This is an arrow of such; it shows the direction and points at which metal will corrode. The intensity of this corrosion is determined by the current density, the amount of current that is flowing from one metal to the next. The value of the potential does not take into consideration other factors that have a far greater effect on corrosion. One of the most important in determining the expected intensity of galvanic corrosion is the ratio of areas of the two dissimilar metals. This will have a significant impact on the current density that can be expected to occur.

GALVANIC CORROSION AND THE RATIO OF AREAS

If the area of the cathode is much greater than the area of the anode, the current density is larger, and this voltage potential difference becomes a critical factor. This is known as the ratio of areas. Galvanic corrosion is dependent on the relationship of area of contact or more specifically the ratio of area of cathode to area of anode. The larger the cathode, the greater the galvanic current and the smaller the area of cathode, the less current will flow. For example, if a stainless steel fastener is used to fix a large steel cross section or panel to a steel frame, the area of stainless steel in relationship to the area of steel is very small. The galvanic current will be constant between the two metals but the corrosion per unit of area of the steel is low because the current density is low. Stainless steel is more noble than the steel and the area of steel is significant in relationship to the small stainless steel fastener. The stainless steel will be protected by the larger steel anode and corrosion of the fastener, even when exposed to a strong electrolyte, will be negligible if at all. At the same time, the steel will experience very little corrosion if any at all, because of vast differences in mass. The opposite condition occurs where the mass of stainless steel is large compared to the mass of a more active metal, such as a steel fastener in a stainless steel assembly, and steel can then experience rapid corrosion.

This is one of Faradays laws where a given current passes between the anode and cathode in a galvanic cell at a proportional rate. If, for example, the cathode area is 10 times larger than the anodic metal area, then the current will be 10 times as great passing through the anodic metal and corrosion of the anode will be rapid.

If the conditions are such that you are concerned with the possible detrimental effects of steel coupled with another metal, then coat the more noble metal with an electrical insulating substance. If the more active metal is coated and there is a breach in the coating where the least noble metal is exposed, then the ratio of areas goes up. There will be a small area of anode to a larger cathodic region, which is the inverse of what you want to have. You want this value to be very small. By coating the cathode, even if the coating is breached, the exposed surface area is a fraction of the anode surface area.

$$\text{Ratio of Areas} = \frac{\text{Cathode Surface Area}}{\text{Anode Surface Area}}$$

Possibly the most damaging galvanic condition that arises and that can lead to other corrosion issues is the degradation of steel substructures in contact with stainless steel in a humid or corrosive exposure. Here, the stainless steel is the more noble and can accelerate the decay of the steel at the interface of the two materials. As the steel corrodes, it deposits iron oxide on the surface of the stainless steel. These can over time interact with the stainless steel surface and create small corrosion cells known as pits. Pitting corrosion is an autogenic process that can inflict itself on stainless steel surfaces, particularly in high-chloride environments. The galvanic corrosion of one can lead to, at best case, red oxide staining and, at worst case, initiate pitting corrosion of the more noble stainless steel.

KEY CONDITIONS NEEDED FOR GALVANIC CORROSION TO OCCUR

For galvanic corrosion to occur, several key conditions must be met. Eliminating any one of these conditions and galvanic corrosion will not happen. Oxygen must be available to the cathode, the more noble metal. This usually is supplied from the electrolyte as dissolved oxygen. In the absence of oxygen, hydrogen ions form at the cathode. This can create a different issue with cathodic corrosion but in the context of art and architectural surfaces, oxygen is usually available.

There must be an electrolyte that provides the connection of the circuit. If the environment is dry, the electrolyte is eliminated, and galvanic corrosion will not occur. If there is a barrier such as a coating of nonconductive material or nonconductive oil, the flow within an electrolyte is prevented.

There must be electrical flow from one metal the anode to the cathode. This is where the two metals are in contact. This is dependent on the difference in electrical potential.

All metals have an electrical potential as shown in the electromotive chart, Figure 7.41. The electrical potential is the tendency of one metal to give up electrons when submerged in an electrolyte. The electrical potential is a measure of a particular metal submerged in an electrolyte and compared to the electrical potential of a known electrode of another metal. It is an indicator "arrow" showing the electrical relationship of metals. If two metals are close to each other, the electrical difference should be less and thus the rate of corrosion will be slower. Factors such as current density determined by the ration of areas are more important than the relationship of the metal on the chart.

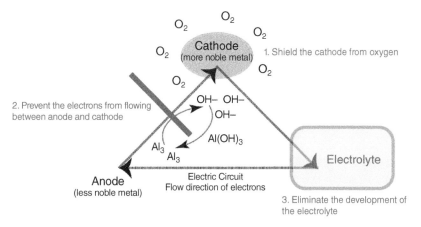

FIGURE 7.41   Galvanic circuit and means to prevent it from occurring.

Key Conditions for Galvanic Corrosion to Occur

- Oxygen available to the cathode
- Electrolyte
- Flow of electrons

TRANSFER CORROSION

Contamination of steel surfaces from copper particles is rare. Typically, it is the reverse. A steel assembly, most likely painted, is above the copper surface of a roof or enclosure. The steel corrodes and the particles of rust are deposited onto the copper surface, creating a stain.

Transfer of iron particles to stainless steel in the manufacturing and handling process can create galvanic cells on the surface that can lead to rust stains, or even worse, pitting. Contamination from steel particles becoming embedded in the surface of the stainless steel is a common problem. For instance, grinding of carbon steel in an area near stainless steel, brass, copper, or aluminum parts can send hot, semi-molten particles of steel onto the surface of these metals.

Steel fabrication processes rarely protect steel sheet or plate with PVC films or paper interleaf. Steel particles can come off onto the cutting surfaces and forming dies used to fabricate steel assemblies. If the tooling is subsequently used on stainless steel fabrication, particles can be transferred to the unprotected stainless steel or copper surface.

To prevent transfer corrosion, the most basic approach is to not mix manufacturing processes that work with both carbon steel and other metals. Grinding discs, cutting tools, cutting tables, and glass bead blasting equipment all should be devoted to steel and not intermixed with other metal fabrication. The facility itself should have distinctive, separate areas for working on carbon steel and working on other metals.

This is not the cast with weathering steels. Contamination from low-carbon steel will intermix with the growing oxide on the surface of the weathering steel and will not harm it or the subsequent final oxide development.

PITTING CORROSION

Steels will develop pits from exposures to corrosive substances.

Concerns often are when the pits are hidden and form from the backside of the surface where it remains moist and the pit develops and grows until the metal is perforated.

Pitting corrosion is defined as "the localized dissolution of an oxide in the presence of a solution containing certain anionic species." Chloride ions are anionic with their added electron making them negatively charged.

Pitting corrosion is sporadic across the surface of steel. It can be localized on the surface but usually exhibits a stochastic behavior rather than a uniform attack on the surface. The problem with steels is it can happen on either side and is self-perpetuating.

Pitting can be initiated by deicing salts, fertilizer, and other particulate matter, even small steel particles from other sources. These cause a localized degradation as moisture is added, setting up small corrosion cells. Weathering steels can experience pitting corrosion as small discolored spots form on the surface. Figure 7.42 shows a steel plate surface that has pitting. The left image is a close up of the surface. The right images show the remnants of the pitting after blasting. Remnants of the discolored regions remain.

INTERGRANULAR CORROSION

Intergranular corrosion can occur with any metal made of contrasting elements. The areas around the edges of a grain that develops in a different orientation to other grains are sites where intergranular corrosion will occur. For the exposed steels used in art and architecture, this can occur, but it would tend to be overwhelmed by other corrosion conditions first. This is not a particular concern for weathering steels used in art and architecture.

CREVICE CORROSION

Steel is susceptible to a corrosion behavior referred to as crevice corrosion. Crevice corrosion can occur where small gaps, less than 2 mm, between surfaces of steel meet or where surfaces of steel are covered with gasketing or other semi-porous material. Even loose paint that traps moisture can create crevice corrosion. Crevice corrosion, by its nature, is insidious. It is concealed from view. It does not extend out from the crevice and can only be uncovered by disassembly.

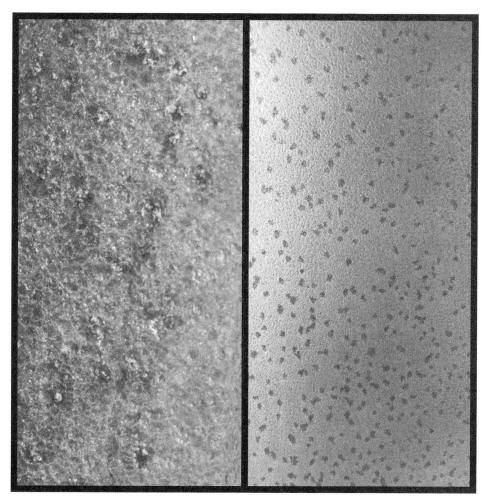

FIGURE 7.42 Example of pitting corrosion on a steel surface.
Source: L. William Zahner.

For crevice corrosion to occur on steel, moisture is all that is needed.

Crevice corrosion is essentially a design and maintenance issue. Well-designed joints can prevent the occurrence of crevice corrosion. One can eliminate the crevice by using elastic materials that allow thermal movement without bunching up, using materials that do not absorb water and designing well-drained joints that allow rains or maintenance to flush the joints. If joints cannot be avoided, consider making them larger so they can be adequately flushed out and inspected. Figure 7.43 shows a cast iron window frame and wrought iron ornamental rail. The chipping paint is trapping moisture as well as the tight bends on the rail. Gaps in the seal around the glass hold moisture on the cast iron window frame and cause the paint to spall.

FIGURE 7.43   Wrought iron rail section showing corrosion on cracks in paint and tight recesses.
Source: L. William Zahner.

LINE CORROSION

When metals are partially immersed in one solution while parts are exposed to another solution, an electrochemical polarity can form—in particular, if one solution is more oxygenated than another. For art and architecture, this might mean a fountain or basin region that retains water, or it might involve partial burial in soil. The area of interface can be more prone to corrosive attack. The more aerated region becomes cathodic relative to the less aerated region. Oxidation and possibly pitting can become more prevalent at the interface region. If the weathering steel is to be partially buried or immersed in water, consider coating the steel with a barrier of thick bituminous paint.

The simplified Pourbaix diagram, Figure 7.44, provides a quick visualization of the oxidizing and reducing abilities of stable compounds of iron and iron oxides. Essentially, at any given point on the diagram, the stable compound, thermodynamically, is indicated for a given pH and E, the electrochemical voltage.

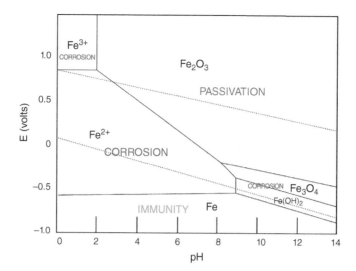

FIGURE 7.44 Simplified Pourbaux diagram for iron.

The pH alone is not a clear indicator of what the iron will undergo in moist exposures. The oxidizing potential and the reducing potential of the particular environment are needed. The oxidizing criteria go up in well-aerated environments and the reduction potential increases in polluted environments. This oxidation potential is represented by the electrochemical potential of the solution.

From a quick overview of the diagram, you can see that Fe_2O_3, iron (III) oxide, is passive over a broad range of exposures. This is the red-brown deep layer of oxide that forms on weather steels.

CHAPTER 8

Maintaining the Steel Surface

Lay not up for yourselves treasures upon earth, where moth and rust does corrupt . . .

Matt 6:19

INTRODUCTION

Steel surfaces want so badly to change back to their mineral constituents. The least exposure to moisture, and the bare surface of steel starts looking for oxygen. This puts the use of steel by the designer in a precarious position. We see the metal for strength, durability, and hardness, but as long as mankind has interfaced with iron, we have had to accommodate this tendency to corrode.

The various exposures encountered by steels used in art and architecture include interior surfaces or the extensive variations encountered in the external environment. Near the seacoast, urban environments, rural exposures, or those exposures that fall in between, each pose a unique challenge to the use of steels.

To be clear, most steel and iron fabrications are coated, either with paint, lacquers, or some other metal such as zinc (galvanizing) or chrome. The weathering steels offer a different, engineered approach to the environment. We will touch on the maintenance suggestions for these various surface coatings as they relate to the steel base metal but will concentrate more on the exposures of weathering steel surfaces in our built environment.

PAINTED STEEL SURFACES

For the painted surfaces, a basic maintenance program should be agreed upon by the designer and the end user, the client. Too often, steel fabrications used in architecture and art are left to a "wait and see" approach. There are few areas where this approach makes sense. We periodically clean our vehicles, removing road grime and deicing salts. Why not consider this for our sculpture, building entryways, and painted steel surfaces? Natural rains go a long way in many applications where design considerations have eliminated ponding water and hidden recesses where moisture can be trapped. But unforeseen, unique conditions are always present in the world. Things change, and time waits for no one.

For the painted surfaces, the initial rust bleeding out onto the paint surface or showing through the paint should be a call to attention, a warning flag that says something needs attention. See Figure 8.1.

FIGURE 8.1   Rust appearing on painted surfaces.
Source: L. William Zahner.

This initial rust is iron (III) hydroxide, knows also as ferric oxyhydroxide. This is the orange red rust that runs and streaks. It can stain some porous paint surfaces, but for most high-quality paints it can be removed from the paint surface without harming the underlying paint. Mild phosphoric acid and commercial rust removers are not solvent based and will not dissolve the organic coatings that make up paint. These are dilute acids and will dissolve the rust, allowing it to be removed by rinsing or wiping. These solutions require the temperature to be above 15 °C (60 °F) to be effective and, depending on the amount of rust, may require several treatments. It is very important to rinse these acids from the surface after treatment.

Rust Removers

- Dilute phosphoric acid
- Dilute citric acid
- Dilute oxalic acid

Any of these acids should not harm the paint coating. Eye and skin protection are warranted when using acids. Once the rust has been removed, thoroughly rinse the surface allow to dry and treat the area where the rust had appeared.

The underlying source of this corrosion can be more difficult to arrest. Seams that allow moisture to enter or collect are very difficult to contend with. You can remove the stains from corrosion that have collected but the source may be hidden. The source is most definitely exposed base metal that is coming in contact with moisture. The various corrosion products shown in Figure 8.1 are coming from other regions, and it is difficult to be certain how to stop the continued deterioration of the base steel.

Several techniques are possible after all the visible corrosion has been removed and the surface is clean, dry and all loose paint particles removed:

- Apply a zinc-rich coating into and around the region.
- Spray as far into the gaps between the metal surfaces and coat the exposed base steel regions with a zinc-rich primer.
- A pigmented barrier top paint layer can be applied.

Once everything has dried completely, seal all regions where water can enter the space. Use a compatible sealant, but one that can be removed. Inspect the area periodically to ensure the corrosion has been arrested. Remove a portion of the sealant and see if rust has formed or trapped moisture has occurred. If this is the case, you will need to reclean the rust, allow to dry and reapply some of the zinc-rich paint. Look for the entry point for the moisture and seal it.

DEVELOP A MAINTENANCE STRATEGY

Cleaning and Maintenance – Weathering Steel

Cleaning and maintaining weathering steel surfaces are not overly difficult. Similar to other coarse surfaces, dislodging staining materials that have keyed into the layers of oxide without damaging the oxide is the main task.

Avoid using acids or acidic substances like vinegar. These can harm the oxide coating. Industrial grade detergents, alcohol, organic cleaners such as lacquer thinners, xylene, toluene, and acetone will not affect the oxide surface.

Pressure washing the surface can remove loose particles and debris without damaging well-established oxidation. The oxide is very hard and tenacious when formed and will withstand 130 bars (2000 psi) when sprayed from a short distance. Test an area to be certain the oxide has sufficiently adhered before blasting the overall surface. Use clean water or water and detergent. Do not use abrasive. Schedule a thorough rinse of the surface several times a year to remove debris and any bird waste. See Figure 8.2.

A periodic wash with clean water and a mild detergent followed by a rinse in clean water performed once or twice a year will go a long way in keeping the surface of a painted object from corroding because of surface pollution that may be on the paint. If deionized water is available for the rinse water, it should be used. Deionized water is water that has had the ions removed. Different than distilled water, deionized water will capture the free ions on the surface, particularly salts, and hold them so they can be rinsed from the surface.

A light scrubbing to remove road dirt, animal waste, and other detritus that may have become adhered to the surface may be needed. When scrubbing the surface use a nonabrasive material such as a bristle brush or light nylon brush. You want to avoid scratching the surface and thinning the oxide coating.

Steam cleaning with portable, pressurized steel cleaners is an excellent method of removing most soils and even bird waste. The heat and humidity of the steam aid in growing the oxide as well. Cleaning a surface with steam is an excellent method to consider when maintaining a weathering steel surface.

During this light cleaning, an up-close inspection of the surface can be performed to identify areas where corrosion is showing through the coating. Areas of ponding and weakness in coating performance can be noted. Keeping a maintenance log with images is an excellent method to identify areas of concern and monitor any changes. This way, strategies can be devised to reduce ponding or address the weaknesses before they lead to extensive damage.

Maintaining a Steel Surface

- Periodic wash with clean water and mild detergent.
- Light scrubbing to remove stubborn waste products.
- Inspect the surface, photograph, and log any issues for review.
- Develop a maintenance strategy.

FIGURE 8.2   Bird waste on the surface.
Source: L. William Zahner.

Once a paint coating begins to chalk, fade, and chip, there is little you can do short of refinishing. Chalking is the development of a dull, whitish powdery appearance, thus the comparison to chalk. This is caused by the polymer within the paint decaying from ultraviolet radiation. Essentially, the chemical bonds that link the elements of the polymer chain begin to break. Fading is the degradation of the pigment in the polymer. Usually ultraviolet light absorption will induce eventual fading, but chemical interaction can also cause pigments to fade. Once this occurs, the color will be lost from the surface. Sometimes you can add a higher-gloss clear coating that enhances what pigment remains and bring out the color. This is only a temporary fix.

Chipping or flaking of the paint is due to failure of the adhesion of the paint to the base metal surface. This could be caused by thermal changes and differences in elasticity or loss of elasticity from ultraviolet degradation. You can very carefully touch up chipped or flaked regions but touchup paint will act only as a temporary barrier. Color matching faded pigments will be challenging, and the surface may need to be roughened to improve adhesion of the new paint. This roughening can create a reflective difference that is visible at different angles, often making the fix worse than the original issue. Another issue with touchup paints is that they are comprised of a different polymer, usually one of lower quality. As they are exposed to the environment, they will fade at different rates. An overzealous effort at touching up can lead to a spotty surface as exposure to the sunlight over time degrades the touchup paint. It is important to use touchup paint sparingly. If the surface has a scratch, then apply the touchup into the scratch and remove all excess paint.

Protecting Steels – Storage

Use only distilled or deionized water as a rinse to clean the surface of the steels. Dry them thoroughly afterward. Keep the temperature and relative humidity constant and low. For every 10 °C increase in temperature, the corrosion rate doubles.

Keep the surfaces dry and, if possible, apply a thin coating of rust inhibitive oil to the surface. There are several proprietary antirust oils available. These are temporary treatments that can be applied via roller or spray. They can be removed after the steel has been fabricated and before the darkening, paint coating or preweathering.

Protecting Steel Surfaces – Custom

Chapters 3 and 7 discuss the application of conversion coatings on carbon steels to produce certain aesthetic and corrosion resistant characteristics. Theses conversion coatings are widely used in other industries and involve creating a coating that inhibits ionic diffusion out of the base metal and into the base metal.

These coatings are often used to prepare a surface for paint application. These treatments include zinc chromate pretreatments and phosphate pretreatments.

The firearm industry has several proprietary systems that induce phosphates onto the surface of carbon steels. These are zinc or magnesium phosphates. They provide a level of protection to the steel surface as well as an appealing appearance. Gray or black colors with a matte color are the typical coating tones induced into the steel. These are excellent coatings for retarding the movement of moisture to the steel surface and reducing the oxidation – reduction behavior of mild electrolyte effect on the steel. They are not readily applied to sheets or plates and they are crystalline, so they will crack if post formed.

Protecting Steel Objects – Impressed Current Protection

This system of protecting steel objects comes from methods used to protect towers and other difficult to coat structures. It is sometimes considered for protecting sculpture. The process is also known as anodic protection. It works for steel and titanium assemblies and objects. The concept is to induce a passive behavior to the surface by anodic polarization. A slight current is applied to the steel object, enough to overcome a critical current level and induce a nonreactive surface. The current is monitored to keep the steel part in a range where the surface does not what to join with oxygen and remains passive. This process works with for a large surface area but will not protect against chloride ions. It does work well with normal environmental exposures.

Protecting Steel Surfaces – Sacrificial Anode

Steels can be protected by sacrificial anodes, which will provide cathodic protection. This method includes the localized protection provided by galvanizing (zinc coating) and by zinc–aluminum coatings. This form of protection does not need a power supply but works on the phenomena of cathodic protection where one metal, in this case zinc, will sacrifice cathodically for the more noble metal steel. Aluminum will as well, but to a lesser degree. Aluminum coatings on steel provide a good barrier due to the aluminum oxide that forms, whereas zinc affords a level of proximity protection. Cut edges and scratches through the zinc coating "self-heal" in the metal sense by extending protection across a small distance. The zinc oxidizes in place of the underlying steel.

Occasionally, when using galvanized steel for exposed structures or as exposed surfacing panels, the galvanized surface is damaged. Particularly on large, heavy sections, handling and stacking can damage the soft zinc coating. The approach often taken is to apply a coating of zinc-rich paint on the damage. Initially, the appearance is satisfactory, but after a short period of exposure, the paint darkens in comparison to the hot-dipped galvanized surface. It is better to remove all excess paint or not to use it at all.

Zinc coatings eventually will wear as they are consumed by the oxygen-reduction process involved with protecting the base metal. The underlying steel will begin to corrode as the surrounding zinc continues to oxidize. The initial appearance is spotty but eventually it broadens out. Figure 8.3 shows a galvanized form set on a pier in Seattle. The galvanized held up for the first 12 years and now, after nearly 25 years, the zinc coating protecting the steel is nearly spent.

The time it takes for corrosion to appear depends on the thickness of the coating and the environmental exposure. You should expect 11 to as much as 30 years on hot-dipped galvanized coatings. Appendix G shows various thicknesses of zinc applied by means of immersing the steel in a bath of molten zinc.

Once the zinc gives way and the base steel begins to corrode, the only solutions are to recycle the form or paint the surface. You will need to remove the corrosion first. If the corrosion is severe, there will not be sufficient steel to coat with paint.

MAINTAINING STEELS IN DIFFERENT ENVIRONMENTS

Weathering steel is the only steel used in art and architecture where the surface is expected to develop an oxide and resist change from environmental exposure. The oxide is an integral extension of the steel and offers long-term and effective protection from what typically would occur in the environment.

A designer has two choices: natural weathering of the surface as the steel is exposed to atmospheric moisture and sunlight, and the accelerated preweathering or artificial weathering of the

FIGURE 8.3   Ron Fischer sculpture in Seattle.
Source: L. William Zahner.

steel. These were described in Chapter 3. In both cases, the ultimate goal is to achieve a sound, inert oxide coating on the surface in a reasonable amount of time.

To achieve this sound oxide layer, it is crucial that we clean the surface of the steel. Once clean, and the surface develops the oxide, it is important to maintain a marginal level of cleanliness.

For the coated steel surfaces, this boils down to how well the paint coating performs and if it has undergone deterioration. This book has emphasized what happens to weathering steel surfaces. These are the natural, uncoated oxides. For weathering steels, we will look at the surface condition before and after the oxide has been produced.

There are several categories that classify a metal surface for cleanliness. Each of these establish the base line as the metal and the form of the metal as it was intended. The categories are:

- Physical cleanliness
- Chemical cleanliness
- Mechanical cleanliness

The following define the basis of these categories in the context of maintaining a surface for art and architecture.

Physical Cleanliness

Physical cleanliness relates to the degree the steel surface, is free of grease, oils, polishing compounds, and other soils that are not related to the metal but can be found on the surface and have an altering effect on the development of the oxide. Fingerprints would fall into this category as would bird waste, adhesives and other foreign substances that detract from the aesthetic and can lead to premature and unplanned changes to the surface. Figure 8.4 shows stains from bird waste and stains from oil spilled onto the surface. If these are not removed, they will alter the oxidation that develops.

FIGURE 8.4   Oil on weathering steel surface.
Source: L. William Zahner.

The oils will inhibit oxidation while the bird waste may intermix organic substances with the porous surface of the oxide.

Physical cleanliness is achieved by degreasing applications that dissolve and displace these organic categories of foreign matter that find their way onto the surface, either before or after the oxide has developed. Degreasing can involve solvents or solvent vapors when greases or oils are on the surface. It is generally not practical to immerse large assemblies into an electrochemical bath or solvent bath that removes organic soils from the surface of the metal. Milder solvents applied under special controls in order to capture the effluent can be used. They require proper safety equipment and procedures and must follow proper disposal procedures to avoid all impacts to humans and the environment.

For most art and architectural uses, physical cleanliness resides in less industrial cleaners and more biodegradable or readily accessible cleaners such as deionized water, isopropyl alcohol, and mild detergents.

Achieving Physical Cleanliness

There are a number of soils that can find their way to the surface of steel fabrications. On certain finishes, they are not always that simple to remove. On most painted surfaces, unless they have keyed into the paint itself, they are more readily removed. If, however the steel is galvanized or if the steel is fresh from the mill source, oils may be present. When the oils have been removed, the surfaces will fingerprint quickly as the porous surface captures the oils from your hands.

Mill Oils

Mill oils are light-grade hydrocarbon oils applied at the mill after the final cold-rolling pass. These oils are used to combat oxidation during storage and handling. If the steel is destined to weather naturally, these thin oils will evaporate or rinse from the surface at the first rain. These oils are present on steels that are uncoated. Galvanized steel has the oil removed from the steel as a process step before hot dipping.

Other carbon steels often arrive in a "pickled and oiled" condition. They can be ordered dry, but if the quantity is small, they usually have some mill oil protection.

For accelerating the weathering of steel surfaces, these oils must first be removed. They can be rinsed from the surface, removed with steam, or displaced from the surface using a mild detergent.

Fingerprints

The first time the new steel surface is touched by the bare hand, a fingerprint can be seen. The most minor touch can leave a very thin film of human perspiration on the surface. This thin film interferes with light reflecting off the surface, giving rise to a darkened smudge mark. The difficulty is that it remains on the surface until it is displaced with another fluid or it is removed.

FIGURE 8.5   Fingerprint on blackened steel wall.
Source: L. William Zahner.

Fingerprints can etch galvanized steel and can affect the unoiled conversion coated steels if they are allowed to remain on the surface. Fingerprinting is a challenge for many metals used without clear coating or sealing treatments. Figure 8.5 shows an interior darkened steel surface that has been touched repeatedly showing a reddish smear. This did not develop immediately, but over several months the fingerprints and handprints appeared. These cannot be removed with water and detergent. They will require a mild rust remover to effectively remove the oxide. If the coating protecting the conversion coating is still sound, then the rust only will be removed but it will return if not resealed.

Fingerprints are composed of organic oils and fats called lipids, amino acids, and water produced by the body. There are often salts intermixed in the oils, but the amount of chloride in a fingerprint is insignificant. The water will evaporate but the oils and fatty substances remain. It is these oils and fats that are so persistent and difficult to remove. Many cleaners just move the oils around or thin them out. The film from our hands is a light organic oil, and this very thin deposit enters the minute ridges and pores on the surface. Similar to aluminum or copper, steel will be etched by these oils.

Interior steel surfaces of weathering steel will show fingerprints and oils more readily than steel surfaces that have been exposed to the atmosphere for a period of time. Steel surfaces have an affinity for light oils. The oils penetrate deep into the surface pores of the metal. Weathering steel surfaces exposed to the environment for a period of time absorb water from the air. This water fills the pores, effectively preventing light oils from fixing to the surface.

Removing Handprints and Light Soils

- Wipe with clean cotton cloth and 99% isopropyl alcohol.
- Follow with a mild detergent or commercial glass cleaner.
- Wipe with clean cloth and deionized water.

The use of 99% isopropyl alcohol breaks down the oils and smears and gets into the fine grains and displaces the oils. It dries and evaporates quickly, but before it dries out, it has moved the oils closer to the surface. The glass cleaner has a surfactant that lifts and displaces the oils from the surface. Additional wiping down with deionized water will remove any detergent left from the glass cleaner.

On painted surfaces, fingerprinting can be removed easily with commercial glass cleaners and mild soap and water. Deionized water followed by a thorough wipe-down with a clean cotton cloth is an excellent way to remove light fingerprinting.

Galvanized and zinc-aluminum coated steel surfaces can be a little more daunting. Glass cleaners and detergent should be tried first. But often the fingerprint has etched into the zinc coating, requiring chemical dissolution of the oxidized zinc. Figure 8.6 shows fingerprints on a galvanized surface. These have etched into the metal and cannot be removed with organic solvents, detergent, or water.

Removing oils and grease from the surface of steel prior to darkening or weathering is required. It can be as simple as a wipe-down with acetone, xylene, isopropyl alcohol, methyl alcohols, or mineral spirits. Rarely in architectural and artwork preparation is it necessary to go to elaborate hot degreasing or vapor degreasing. Some of these solvents can leave a film when they evaporate from the surface. Remove the film by a final rinse with deionized water or mild detergent.

Oils and lubricants come into play if surfaces or castings undergo some level of machining. Lubricants are also used in deep drawing and some stamping operations. Factories that employ these operations usually have degreasing stations in place. They often use proprietary solutions to remove the lubricants while reducing the environmental and safety exposure these solvents introduce.

Degreasing Using Hot Alkaline Baths

Hot alkaline baths and sprays are employed to remove grease, oils, and lubricants on major production pieces. Typically, hot dilute solutions of trisodium phosphate, sodium meta-silicates, and other alkaline cleaners with proprietary additives are utilized. These require careful rinsing and drying

FIGURE 8.6   Fingerprints etched into a galvanized steel surface.
Source: L. William Zahner.

operations and have limited use in parts and assemblies used in art and architecture. Hot alkaline bath degreasing would be performed in the factory making the steel fabrications.

Dirt and Grime

Weathering steel surfaces and even the hot-dipped galvanized surfaces by their nature are coarse. The oxide that grows on the surface is expansive and crystalline. The microscopic surface is rough and uneven. Dirt and grime can find easy footholds in a weathering steel surface. Vertical surfaces receive the benefit of an occasional rain to remove airborne soils that adhere lightly to the surface. Dew that collects on the surface, however, can also hold the airborne soils in place and redeposit them as streaks, particularly where concentrated drainage creates patterns running down a surface. An occasional rinsing of the surface with clean water and a mild detergent can remove most of these streaks. Adding the power of a pressure-washing pump can aid in the removal. Surface cleaning will also benefit greatly from a final rinse with deionized water. Figure 8.7 shows dirt deposits and stains left as the moisture evaporated from weathering steel. Weathering steel is dark and can get quite warm. This can bake the stains onto the surface, making them more difficult to remove.

Similar to high-pressure washing, steam will effectively remove baked-on contaminants from the surface of weathering steel. The oxide growth will be enriched by the steam cleaning process. Steam gets deep into the metal and pulls out dirt and grime from the surface pores. The energy in the steam acts to break down the bonds that hold the grime and dirt to the surface, and the force of the steam moves the contamination away from the surface.

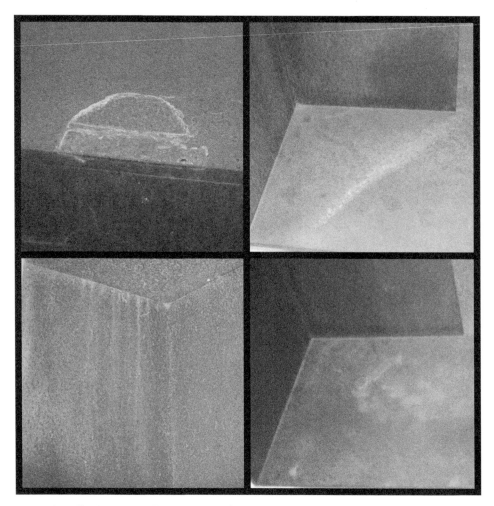

FIGURE 8.7   Dirt collecting on weathering steel surface.
Source: L. William Zahner.

Adhesives

Adhesives and glues on the surface of steel can be removed with solvents. Additions of steam can sometimes upset the gummy substance and aid in the removal from the surface. Solvents such as mineral spirits work well in removing adhesives and glues.

When adhesives remain on the metal for a length of time and bake in the sun, they can change in composition and be very difficult to remove as they become a crust-like, hard substance. This may necessitate a more powerful solvent; ethyl acetate or methyl ethyl ketone (MEK) may need to be used with steam. MEK is considered a hazardous substance. Most hydrocarbon-based

TABLE 8.1   Various types of solvents.

Hydrogen bond donor acceptor solvents
Methyl alcohol
Ethyl alcohol
Isopropyl alcohol
Ethylene glycol
Low hydrogen bonding solvents
Naphtha
Mineral spirits
Toluene
Xylene
Hydrogen bond acceptor solvents
Ethyl acetate
Methyl ethyl ketone
Isopropyl acetate
Butyl acetate

solvents are flammable and require special handling and disposal as well as special personal safety gear.

Table 8.1 shows several solvents and their basic polarity. For example, oxygen in the water molecule is a hydrogen bond acceptor. Different solvents will attract different molecules and break the adhesive bonds that bind them together. It is suggested to start with the top group and work your way down.

Temperature Issues

The dark color of weathering steel makes it a good infrared absorber. Like other steels, it is not a good conductor of heat. The surface exposed to the sun will get very warm to the touch. Moisture on the surface dries more rapidly. Organic material, such as leaves, bird waste, and adhesives on the surface will dry out and can create light staining. Once baked onto the surface, they are very difficult to remove.

These will alter the oxide coating and may require the coating to be removed and restored to bring it back. Solvent wipe to remove the adhesive, saps, and organic oils, followed by steam under pressure, should be attempted first. If these fail to give the desired results, the surface may need to be abraded and the oxide restored.

Deposits from Sealant Decomposition

Silicone and other sealant material are commonly used to close the joints where one material stops and the weathering steel surface begins. It is good practice to avoid the use of exposed sealants on the joints of weathering steel. They collect dirt and stain the adjacent oxidized surface.

Many sealants undergo a catalyzing process as they cure. They contain plasticizers and oily polymers that lead to the development of stains by migrating out of the joint onto the steel surface. The oils are hydrophobic and repel moisture. So, the benefits derived from rains rinsing the soils from the surface is lost around sealant joints. A halo surrounding the joint will appear when the surface is wetted. This is an indication of the presence of the thin oil layer.

Pressure washing with hot detergent alone does not remove the stains from sealants, nor will most mild solvents. MEK will dissolve the stain and remove the soil, but this solvent is not recommended due to the safety and environmental hazard posed by this solvent. Ethyl acetate will aid in the removal of the stain if you allow it to sit and work on the surface. Generally, solvents tend to soften the silicone and urethanes used in sealants and allow them to be wiped from the surface.

There are a few proprietary products that will dissolve silicone. With any system used, personal protective gear along with a thorough rinse-down is a requirement.

Grease Deposits from Building Exhaust Systems

Grease that collects on the steel surface around exhaust systems used to vent kitchens or other food processing can be an unsightly challenge to remove. These hot deposits will darken the coarse steel surface and can enter deep into the surface. This type of stain can be eliminated by proper design and venting. The difficulty lies in the mass of the oily exhaust. Composed of carbon and fatty substances, these organic soils will alight on the steel surfaces and coat the metal with a sticky crust. This greasy deposit is applied warm and can develop into a deep crust containing mold and bacteria as well as the decomposing fats.

Hot steam under pressure with detergent can remove much of this. Deep cleaning with an organic acidic cleaner such as acetic acid or phosphoric acid along with a mild abrasive will aid in dissolving these tough surface deposits. However, these will alter the appearance of the weathering steel in these regions.

Hydrogen peroxide made into a paste with baking soda (sodium bicarbonate) and allowed to set on the surface for a few minutes can loosen the organic soils.

Graffiti

The oxide that develops is thick and dark in color. Most graffiti are dark in color and the contrast just does not benefit the urban "artist" enough. But in the event that a surface is accidentally or purposely painted, the paint can be removed with solvents. Weathering steel will not be affected by the application of any common paint removal solvent. Even strong organic solvents will not affect the oxide or the base metal. The oxides that form on the surface are inorganic, mineralized substances that will not dissolve in solvents used to remove paint. Remove the paint and rinse the surface down.

FIGURE 8.8   Graffiti improperly removed from the surface.
Source: L. William Zahner.

Difficultly may arise on newly formed oxide because it has not yet formed into the tight, strong, crystals on the surface. Wiping the surface can smear the weak oxide and make a simple task of removing the paint, into a difficult task of restoring the surface oxide. The paint will need to be removed and the surface will need to be reworked to bring back the oxide. The area around the region where the paint was removed will require reworking the surface oxide as well to make sure of a match. Figure 8.8 shows a surface that had graffiti removed. Unfortunately, an abrasive pad of some sort was also used, and this smeared the oxide surface.

Eventually the oxide will grow to match the adjoining regions. But this is avoidable. Figure 8.9 shows how the solvent alone will remove the paint without abrasive. This can be used to remove grease pencils and other organic substances applied to the surface of weathering steels.

Graffiti on weathering steel and galvanized steel is removable with minimum aesthetic damage if caught early. Figure 8.9 shows a demonstration where the black paint is partially removed with a wipe of xylene and then rinsed with water.

Graffiti on steel surfaces that have been painted or clear coated pose a different problem. Most spray paints are acrylic. If the base coating used to protect the steel is an acrylic, exercise caution when applying solvents as they can remove the base coat paint. In coil-coated steels or

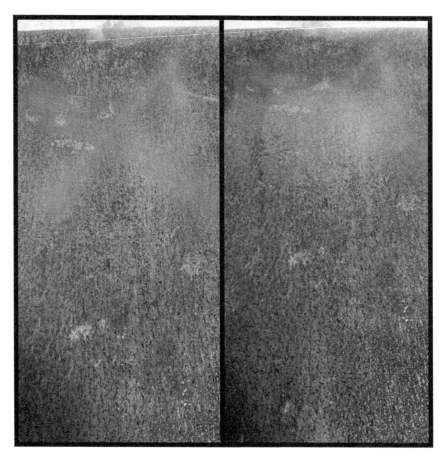

FIGURE 8.9   Removing graffiti paint from a weathering steel surface.
Source: L. William Zahner.

powder-coated steel, graffiti can be removed using naphtha or other mild solvent. Refer to Table 8.2 and try the low hydrogen bond solvents first.

There are a number of tools at one's disposal to achieve a physically clean surface. Most of these tools are readily available. Physical cleanliness considers the removal of foreign substances that have not damaged the underlying metal or metal surface but instead are adhering or bonding to the surface. Removal, then, requires lifting the substance from the metal and rinsing it away. Some of the tools are listed below.

High-Pressure Water Blasting

Similar to washing your car, high-pressure water blasting is an excellent and inexpensive method of removing substances adhering to the steel surface. On well-prepared weathering steel surfaces, ones that have a thick, developed oxide, will not flake when hit with high pressure washing. General

TABLE 8.2 Physical cleanliness.

Condition	Cleaning regimen-physical cleanliness
Oils and grease	Detergent, steam, solvent degreasing
Fingerprints	Detergent, glass cleaner, isopropyl alcohol, deionized water
Dirt and grime	Detergent, steam
Bird waste	High-pressure wash, detergent, steam
Adhesives, gums	Solvent
Silicones	Solvent
Building exhaust	Solvent, steam, detergent
Graffiti	Solvent
Concrete or plaster spatter	High-pressure wash

dirt and bird debris are removed as well as chloride salt deposits before they can have a detrimental effect on the steel surface. High-pressure washing should use clean water, with or without detergent. A final rinse with deionized water if possible is another added means of removing soils and minute deposits from the surface of weathering steel.

Steam Cleaning

There are several commercial steam-cleaning systems that operated similar to high-pressure water blast but instead incorporate a small boiler that heats the clean water up and delivers a blast of hot steam. The pressurized steam gets into the metal pores and breaks the bonds holding the foreign substance to the metal. For weathering steel, steam also promotes the development of the surface oxide.

Deionized Water

Deionized water is an excellent final rinse for metals. Deionized water is water that has had all the dissolved salts removed. Water is passed through a series of tanks containing charged resins, called ion exchange resins. Essentially what occurs is, the metal ions dissolved in the water, calcium, iron, magnesium, sodium, copper are attracted to the anion resin and exchanged for hydrogen ions. The metals are held in the anion tank for later disposal. The anions in the water, chlorides, fluorides, nitrates and sulfates are removed in the next tank containing cation resins and exchanged for hydroxyl ions. The resulting water is free of metal salts. The process of making deionized water is quicker and more economical than distilled water.

The deionized water attracts other minerals and free elements on the surface of the metal by virtue of the water alone. Since the ions have been removed, deionized water seeks out all ionic particles that can be on the surface of the metal.

TABLE 8.3   Scrotchbrite pad and corresponding closest grit produced by finishing.

Color of 3M Scotchbrite Pad	Comparable grit
Tan	120–150
Dark gray	180–220
Brown	280–320
Maroon	320–400
Green	600
Light gray	600–800
White	1200–1500

Abrasive Pads and Brushes

Frequently, physical effort is needed to remove stubborn substances. There are abrasive pads made from nylon impregnated with various cutting compounds. One of the more popular is the Scotchbrite® pad manufactured by 3M. There are other similar pads, each with a comparable grit. See Table 8.3.

Chemical Cleanliness

This relates to a surface that is free of oxides, chlorides, nitrides, and carbides that are not related to the base metal or intended to be part of the developed oxide for weathering steels. In this case, these oxides have formed on the metal by outside additions or unexpected influences. These additions can lead to chemical reactions with the steel surface and pose long-term performance concerns and appearance issues.

Sacrificial materials such as zinc from backup supports or zinc-coated fasteners can develop streaks or stains. Figure 8.10 shows exposed fasteners leaving streaks on a weathering steel surface. These are coated fasteners with a more active coating, most likely zinc or a zinc aluminum. These will remain but darken over slightly.

To be clear, weathering steels, as they develop the thick oxide, are little affected by other oxides, even copper oxides that may drain onto the surface. These other oxides will be incorporated into the thickening iron oxide layer. They may for a while have a negative aesthetic appeal.

For the weathering steels, it is more often the case where the more soluble iron oxide develops, sluffs off, and stains adjacent materials. This can happen with steels as well. Figure 7.2 shows steel particles depositing on a copper roof.

When preparing the carbon steels for darkening or paint coating, the oxides and scales that may be on the surface will need to be removed or they will interfere with the development and

FIGURE 8.10   Stains from fastener corrosion.
Source: L. William Zahner.

paint adhesion. The development of the weathering steel oxide during the preweathering process requires the surface to be clean. The oxides and scale should be removed completely. If these remain on the surface, they will impede the oxide growth.

Abrading the steel surface using sand, steel shot, or other abrasive media will remove much of the surface oxide and can, if the pressure power is sufficient, remove scale. Figure 8.11 shows a steel plate entering the blasting chamber. Note the oxides and scale on the surface in the top image. The lower image is after blasting as the plate exits the chamber. The scale and oxide have been removed. Note that once the heavy scale has been removed, there is a "scar" remaining. Scale takes up volume on the surface. Under the scale, the surface is rough.

Chemical cleaning may involve immersing the steel in pickling baths to remove and dissolve oxides and scale from the surface. This renders the surface of the metal back to the original makeup and allows the steel to develop the darkening transition layer or to accept organic or inorganic coatings. This cleaning process must be proceeded by physical cleaning to remove all grease and oils from the surface.

FIGURE 8.11   Scale and oxide removal.
Source: L. William Zahner.

Immersion in pickling baths proceeds the galvanizing process. Once the steel part or plate is degreased, immersion in a bath of acid follows. The acid bath removes all oxides and scale from the surface of the metal.

Laser Ablation

Laser ablation is a cleaning method that can be used on steel to remove oxides, chlorides, dirt, grease, paint, and grime. The advent of the fiber laser development has made this technique a viable form of cleaning and restoring a metal surface to its original form. Laser ablation removes material from the surface of steel by thermal shock and vaporization.

Essentially, the energy of the laser light is absorbed by the surface contaminants, and this absorption of high energy vaporizes the substances without damaging or effecting the mechanical properties of the steel. The laser beam is pulsed to the surface and a tremendous amount of energy is released in a short period of time. This energy is absorbed by the particles on the surface of the metal, even the oils of fingerprints. The particles become excited and vibrate as the bonds between the substances and the steel surface are broken. A flowing gas sweeps the particles from the surface and a suction pulls the particles into a collection filter. In this way, the surface contamination and oxides are collected and disposed of without release to the atmosphere or redeposited on the ground. This also protects the lens of the laser device.

In laser ablation, the laser itself is kept remote and the beam of light is transferred via a fiber cable. The cable length can be more than several hundred feet if necessary, without affecting the power at the beam to metal contact. The portability afforded by the fiber laser allows for cleaning remote and difficult regions in situ.

The most common fiber laser used for laser cleaning and ablation is a Nd:YAG. The Nd:YAG is more efficient than a CO_2 laser and they are easy to handle and operate. The Nd:YAG is a solid state laser based on the rare earth doped crystals of yttrium aluminum garnet, (YAG). These are manmade crystals specifically designed for lasers. The Nd:YAG uses a doped fiber to deliver the high power energy to the laser head. For ablation cleaning the Nd:YAG is normally a pulse laser with an emission wavelength of 1064 nm. There are several alternatives to the Nd:YAG that utilize other rare-earth elements and the development of these lasers is advancing rapidly. But the majority of ablation operations are performed with the Nd:YAG fiber system.

For laser ablation to work, the material's wavelength absorption must be compatible with the laser wavelength. The energy needed to break the bonds of the oxides to the surface and the detritus adhesion is below the threshold of the of the bonds that hold the metal together. When the bonds that hold the contamination to the steel are broken the contamination vibrates off of the surface from the excited, high-energy electrons.

Unlike other methods of cleaning, no waste is released to the environment, the process requires no solvents, applies no water and leaves the surface free and clean of all contamination. Laser ablation can be used on wet or dry surfaces (Figure 8.12).

Special training and care need to be exercised by laser ablation cleaning of steel surfaces. The skill and expense are higher for this cleaning technique than other chemical techniques. However, the residual waste and environmental impact are significantly lower with this cleaning technique.

When using laser ablation, the surface of the steel is altered. The pulsing laser leaves behind a fine etched finish that is different than other mechanical finishes. It can be performed robotically or manually.

Constant Wetting

Weathering steel, like most metals, does not perform well if it is constantly wetted or in an environment that restricts the development of the oxide layer. Constant wetting of the surface keeps the insoluble, oxyhydroxide layer from developing. The ferric oxide sluffs off as it forms just off the surface of the base steel. It must be able to have time to develop the oxyhydroxide layer, which requires periods of dryness. Weathering steel surfaces must be well drained and not act as basins that hold water. This can also damage or impede the oxide layer from growing properly.

FIGURE 8.12   Laser ablation demonstration on an oxidized steel plate.
Source: L. William Zahner.

When weathering steel is embedded in moist soils the oxide will not grow a tight barrier on the steel. Consider coating the steel with a thick, zinc-rich primer or bituminous barrier paint system to protect the surface where it is in contact with the soil.

If the reverse side of a weathering steel surface is not allowed to adequately dry, consider coating this surface with a zinc-rich prime paint or other barrier paint. In particular, when using thin sheet weathering steel, coat the concealed side to prevent the metal from oxidizing on both surfaces. Buried, concealed and submerged weathering steel has been shown to rust away in large sections not unlike normal, unprotected steel parts exposed to moisture.

If the design calls for plates to be overlapped, as shown in Figure 8.13, the reverse side of the steel should be coated. Water will be drawn into this gap and corrode the metal from the reverse side. Water allowed to enter a lapped joint can develop a condition known as *pack rust*. When this is allowed to occur, it can cause a rapid deterioration of the lapping joint from within. This corrosion condition can develop a significant force that is capable of prying the joint apart.

Sometimes, when storing weathering steel outdoors, a similar corrosion condition can develop when sheets or plates are stacked on one another. Moisture allowed to enter between the two surfaces will create a line of thickened corrosion products on the surface of the weathering steel. This corrosion product, if found early, can be removed by lightly sanding and rewetting the surface. It is usually only a superficial layer of loose iron oxide on the surface of the more durable ferric oxyhydroxide below. However, if allowed to remain, the stain can affect the initial surface appearance.

FIGURE 8.13   Shingles of thin weathering steel.
Source: L. William Zahner.

Water Streaks and Standing Water Staining

Water streaks and water stains can be unsightly in the early days of the development of the ferric oxyhydroxide layer. The streaks and stains are of a lighter, contrasting orange color. If the weathering steel is set horizontally and water is allowed to stand on it, it will develop a stain, particularly if there is paper or cardboard protection between the plates and this is allowed to get wet. The paper will hold the moisture to the surface and form variable stains across the surface. Figure 8.14 shows a series of preweathered steel panels that were stored outdoors. They had cardboard interleave that

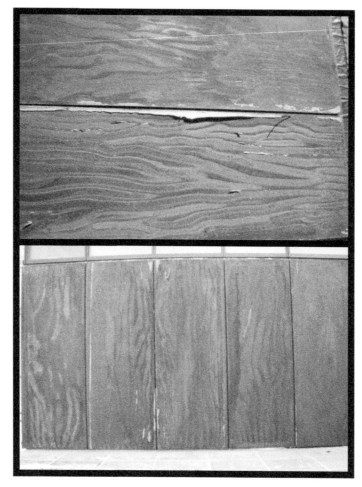

FIGURE 8.14   Stains from saturated cardboard setting on the surface of preweathered steel.
Source: L. William Zahner.

was allowed to also get wet. The result was an interesting stain but not particularly what the client wanted.

Stains such as these can be difficult to remove and will require lightly abrading the surface followed by several cycles of wetting and drying. If you are matching adjoining nonstained weathering steel plates, this can be a challenge. It should be noted the beauty of weathering steel is the natural appearance of the metal. It should not be expected to appear as a painted surface. It derives a natural beauty from the variations on the surface the way wood derives its beauty from the contrasting nature of the grains.

Deicing Salts

Prolonged exposure to deicing salts can damage weathering steel. However, if the salts are rinsed from the surface, the effect will be minor on weathering steel. Figure 8.15 shows a surface of weathering steel that has been exposed to salting around the entryway of a building several times a winter. This exposure has occurred each winter for nearly 40 years. The surface has some minor discoloration identified as the lighter color on the dark background of the deep weathering oxide.

The salts will create discoloration on the surface and can even remove some of the protective oxide layer. These salts should be removed from the surface every spring by thoroughly wetting the

FIGURE 8.15 Deicing salts creating minor discolored oxide stain.
Source: L. William Zahner.

TABLE 8.4   Composition of sea water.

Approximate % by dry weight	Element
55	Chlorine (Cl)
31	Sodium (Na)
8	Sulfur (S)
4	Magnesium (Mg)
1	Calcium (Ca)
1	Potassium (K)

area and rinsing the metal down with fresh water. Weathering steel that has developed a thick, protective oxide layer will resist the corrosive effects of the salts. The surface can be damaged by deicing salts if they are allowed to remain for prolonged periods where they continually are wetted. This draws out iron ions into the solution to form weak oxides. The telltale sign of this is the lighter colored orange rust on the surface caused by the chloride ions interacting and disrupting the trivalent iron. Moisture provides the electrolyte and the ferric oxyhydroxide changes into a ferrous oxide.

In Figure 8.15 the horizontal sill below the metal panel has salts on the surface yet it appears to be holding up very well. It is safe to say that deicing salts will interact with the surface, but the effects in many cases will be minor if the surface is rinsed with fresh waters every spring.

Several chloride salts are used for deicing. Sodium chloride is the main salt deposit in seaside environments (Table 8.4). Whereas in deicing, sodium chloride is usually mixed with other chlorides. Sodium chloride for deicing is usually mined rather than taken from the sea. The mined salt, known as the mineral halide or the common name rock salt, contains numerous other trace compounds. It is the most economical yet the least efficient for deicing purposes.

Deicing salts work by lowering the freezing point of water. Sodium chloride is an endothermic salt, which means it must pull energy from the surroundings. Its effective working temperature is 7 °C (20 °F), so below this temperature it does little good in melting ice.

Another common salt used is calcium chloride. Calcium chloride lowers the working temperature further by creating an exothermic reaction. It releases heat as it goes into solution. Calcium chloride is also hygroscopic, as it can access moisture from the air. Often, the deicing salt is provided in a combination of 60–80% sodium chloride and 20–40% calcium chloride.

Magnesium chloride is being used more frequently in many of the deicing mixes. This salt also reduces the effective working temperature and has less of an environmental impact because it is provided as a hydrate. This compound is composed of more than 50% water. Magnesium chloride is exothermic. As it goes into solution it will generate heat.

Various additives have been used, many of which are organic in nature. These additives make the deicing salts sticky when they are in a slurry form, so they adhere to the road and walk surfaces. They also make the deicing salts stick to metal surfaces. Beet juice is a common organic additive.

TABLE 8.5   Working temperatures for deicing salts.

Salt	Temperature	
Sodium chloride	7 °C	20 °F
Magnesium chloride	−18 °C	0 °F
Calcium chloride	−32 °C	−25 °F

TABLE 8.6   Chemical cleanliness.

Condition	Cleaning regimen-chemical cleanliness
Oxides	Mild acids, such as phosphoric, citric, or oxalic acid solutions. Rinse
Scale	Pickling and abrasive blast systems
Constant wetting	Design. Bituminous paint. Zinc-rich paint.
Water streaks	Lightly abrade and steam
Deicing salts	Clean water rinses every spring. Design.

All of these salts corrode metals when the right conditions are met. Salts, both deicing and coastal, are hydrophilic. They absorb moisture from the air and hold onto it longer. Table 8.5 shows several of the most common salts used and temperature where these salts tend to work to depress the freezing point.

Some alternative deicing treatments do not affect steel surfaces:

- Potassium acetate
- Calcium magnesium acetate

Bridge structures made of weathering steel in northern climates are often attacked by the deicing salts used on the roadways above when these salts develop electrolytes from moisture and condensation and pool on surfaces or enter into joints and seams. This can cause extensive damage to the surface of these structural members. In the United States, approximately 8–10 million tons of salts are spread on the roads and highways each winter.

These are more expensive alternatives to the chloride salts but in the long run, they would reduce the cost of cleaning and maintaining steel as well as improve the immediate environment around the roadways (Table 8.6).

Deicing salts are applied in the cold of winter. The cold temperatures will inhibit and slow chemical reactions way down. When the first warm weather arrives, thoroughly rinse the weathering steel surfaces. This will remove the chlorides and salts before they have the opportunity to develop corrosion cells on the metal. Follow this simple step and the corrosive conditions created by deicing salts will be frustrated.

Mechanical Cleanliness

By mechanical cleanliness, a steel surface must be free of regions that are torn, scratched, dented, and gouged. This would also include surface marring from manufacturing processes and packaging issues that alter the surface. Mechanical cleanliness is a restorative process where the surface regains most if not all of its intended geometry.

VANDALISM

Scratches and Mars

If the graffiti is scratched onto the surface with a rock or key, the finish can usually be restored. This would also include scratches from handling and installation of large sections.

Weathering steel that has develop the oxide is very hard, and most substances will be undermined by the hardness of the weathering steel. However, materials can leave a streak that is light in contrast to the darkened surface. To repair the surface, consider cleaning the surface of loose material. You will need to perform this on more of the surface than simply where the scratch is in order to blend it. Figure 8.16 shows a surface of weathering steel on a memorial that was scratched with gravel. The surface was re-abraded with fine abrasive pads, then rewetted. As the surface is wetted

FIGURE 8.16   Memorial constructed of weathering steel. Initial repair showing discoloring.
Source: L. William Zahner.

down it grows the oxide back slowly, and this should be lightly blended. Eventually, after several wet and dry cycles, it will blend with the adjoining surfaces.

Take a coarse, nonmetallic pad and rub it across the surface, working the scratch and deposited material away. Then wet the surface periodically to even out the oxidized surface. This technique also works well in mild transport damage that may occur to the surface of the weathering steel.

Scratches cannot be removed, but if they are not deep they will quickly oxide and be less conspicuous. Major scuffs on the surface can be treated by first cleaning with a pressure washer. Follow with a light abrasive pad then use steam to bring the surface back to match the surrounding areas. You may need to use a light abrasive pad and work the surface out to the edges. Allow the surface to dry. Repeat the steaming process as needed to help blend the surface and remove the mark.

Dents

Steel and weathering steel are significantly harder and more durable than most other metals in use. For the weathering steels, yield strength is approximately 40% greater as compared to normal steel. It will require a significant amount of energy to dent plates of weathering steel. The oxide that develops is also very hard, as compared to most other surfaces. The initial oxide is weak, but as it grows the oxide gets very hard.

Hailstorms, even the monster-size hail of the Midwest United States, have negligible effect on weathering steel surfaces. They simply do not possess the energy needed to achieve plastic deformation of the metal.

Ghosting from Packaging Materials

On occasion, when steel surfaces are protected and stored for periods of time, the protective media can leave light stains on the surface. In particular, art made from weathering steel that has been wrapped and placed into storage is subject to a surface mark that is a ghosting of the wrap. Moisture can be trapped and held against the weathering steel or the oxide may be rubbed by the packaging material.

To repair the surface, it will be necessary to wet the surface and regenerate the oxide where it has been damaged. Sometimes lightly sanding the surface is needed. Hot steam aids in the redevelopment of the finish surface. To avoid creating a more visual discolored region, the entire surface plane may need to be worked by lightly sanding, steaming, and drying. This will need to be repeated several times until the surface has returned. The far right image of Figure 8.17 shows the surface after restoration.

Distortion

Steels, like other wrought metals, are anisotropic. That is, mechanical properties in one direction are different from mechanical properties in the other. This is due to the microscopic alignment of grains as the metal is produced. Most of the steels used in art and architecture are leveled and stress

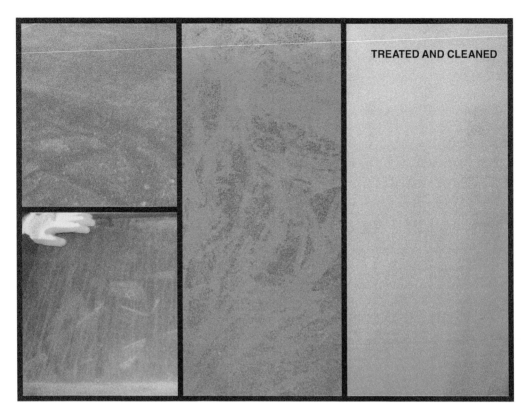

FIGURE 8.17   Ghosting from packaging material.
Source: L. William Zahner.

relieved in such a way that internal directional bias is reduced. Couple this with the low reflective levels of the steels and distortions are not visible. A high gloss paint may, however, reveal distortions, but weathering steel and galvanized steel coatings do not.

Welding Distortion

Distortions created by welding processes are not usually an issue with steel, in particular, weathering steel. The very low gloss and diffuseness of the coarse surface that develop mutes any distortion. Distortions may occur localized around the weld where the metal has drawn in and created a concave distortion. Good welding practices and welding skill will eliminate this from occurring.

Shielded metal arc welding, SMAW, is a popular welding process for plate and heavy sheet but it is slower and will buildup heat lead to distortion. Higher-speed processes should be considered instead, particularly those that are automated such as submerged arc welding, SAW, electron beam welding, and laser welding. Gas tungsten arc welding (GTAW), when automated, will produce good welds in an automated and semiautomated system.

Distortions from Cold-Forming Operations

Weathering steel sheet and plate are provided leveled and flat. The high strength of these alloys requires more power in shearing and forming but typically they appear flat. On horizontal and lightly sloped regions dishing (concavity) of the center should be avoided. Surfaces made from weathering steel should always drain.

Abrasive blasting the surface of weathering steel on one side results in a convexity toward the side being blasted. The surface being blasted is stretched and has more tension; thus dishing toward the abrasive blast will occur. Dishing will occur on thick sheets and plates as well as thin. It is recommended to apply the abrasive blast onto both surfaces to even out the stress and flatten the metal.

The low gloss and coarseness of the surface will conceal minor distortions. They will be measurable, but not visible. The surface is extremely low in reflectivity.

Hail Damage and Small Dent Repair

As mentioned previously, it will take a lot of energy to dent weathering steel plates, or any steel plates, for that matter. Thinner sheet can be more easily dented but rarely by hail stones. Steel used on our automobiles experience hail impact, but this is typically on the horizontal surfaces. The steel used on automobiles is not as strong as the high-strength, low-alloy (HSLA) steels we commonly refer to as weathering steel.

Small dents can be removed. For the dent to have occurred, the steel would have gone through localized plastic deformation. There may be some thinning as the dent stretches the metal slightly, but this is hardly noticeable. If the dent is acute and has a visible crease, the crease will not be removable. If the dents are marks in the surface such as hammer marks, these as well are not repairable. Smooth concave and convex dents are repairable.

To remove a dent, first lightly abrade the reverse side of the panel or sheet. This will pinpoint exactly where the dent is, and it will highlight the area. Block the face side with a clean, smooth, wooden block, end grain against the metal. On the reverse side, the dent can be carefully hammered out with another wood block against the metal. Work it gently and the dented metal will return close to the original position (Table 8.7).

TABLE 8.7 Mechanical cleanliness.

Condition	Cleaning regimen-mechanical cleanliness
Scratches and mars	Light abrasion followed by steam or wet and dry cycles
Dents	Block and hammer to reverse damage. Refinish.
Ghosting from packaging	Fully abrade and restore finish with steam and wet/dry
Distortions	Design and skill

Free Iron Particles – Staining

Weathering steel allowed to weather naturally produces an appreciable amount of rust particles as it develops the more resilient oxide of deep purplish brown. These are the initial, oxyhydroxide and ferrous hydroxide particles of yellowish-orange. They will stain other materials below the surfaces of the steel.

In heavy rains, the water carries the staining oxides away, but dew forming on the surface will collect and run to the lower edges, and gravity will pull these concentrated drips of oxide onto the surrounding surfaces. Even with preweathered steels, there will be some runoff. It is unavoidable that as the oxide develops and thickens there will be porous regions where the soluble ferrous oxide particles will form (Figure 8.18).

The most efficient way to deal with this is by design. Drip edges and catch regions where the rust-laden moisture is directed to, plantings and dark stones along the base are some of the ways that design can accommodate the unappealing stain. Figure 8.19 shows a roof surface draining into a planted region.

FIGURE 8.18   Rust stains on concrete from perforated light bollard.
Source: L. William Zahner.

FIGURE 8.19   Wall draining into stones.
Source: L. William Zahner.

   This oxide is harmless but the aesthetic context of a rust stain on light-colored concrete or stone can be detracting from an otherwise beautiful surface. This staining is the primary reason why preweathered steel is gaining wide acceptance in art and architecture.

   Removing the oxide when it occurs is not easy. There are several commercial cleaners that use trisodium phosphate, citric acid, oxalic acid, and acetic acid (vinegar) in different concentrations. A poultice should be made that sets on the stain and works to lift the iron oxide out of the pores of the surface. You need to apply the poultice to the rust when the temperature is warm, above 15 °C (60 °F). Allow to set on the stain. Use a bristle brush and work the stain up and out of the pores, then thoroughly rinse. On large stained areas, it is best to approach professional metal cleaners. The results may not be complete. It depends on the porosity of the material that was stained. It is also a good idea to test an area to ensure that you do not make the stain worse by abrading the stained material.

Coating Steels

Weathering steel can be painted. If the paint is to provide color and the weathering steel is simply the geometric form, then it should be treated as typical cold-rolled steel where it is blasted to a particular specification, prime painted, and then finish coated.

If, however, the idea is to place a clear coating over the oxidized surface to either protect it or stabilize it to a particular color or tone, then several options are available. Clear coatings can also create a subtle contrast for special effects such as signage.

Clear lacquers are available for metals that should offer performance in most interior applications and limited performance in exterior applications. Other, penetrating type oils that harden like shellacs have shown success. The surface will darken when any clear coating is applied.

For any of the steels, clear coatings as protective barriers are commonly considered.

In rare occasions, it is necessary to coat the weathering steels. It is possible to add additional corrosion protection or to prevent the oxide from being rubbed off the surface and onto clothing.

Clear coatings on interior steels provide various cost to performance and lifespan issues. Figure 8.20 shows the various steels that could be used on an interior surface and cost versus lifespan when some maintenance and restoration is involved. For example, blackened steel with a clear coating will perform and cost less over an 8–12 year period than with a wax coating. Wax may

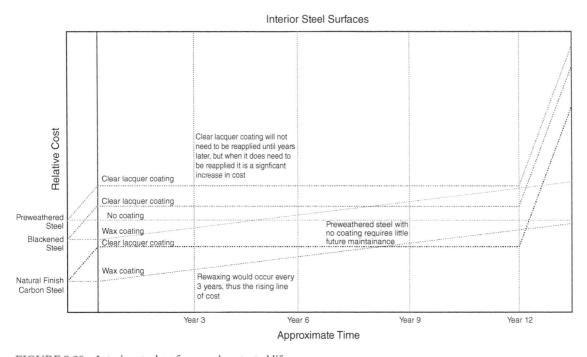

FIGURE 8.20   Interior steel surfaces and protected life span.

be more economical at first, but it requires maintenance to the point it probably costs more, at least until the clear coating has to be replaced. Then the cost shoots up.

The clear coatings used on steels do not perform as well as thermoset pigmented coatings we associate with aluminum and steel parts and assemblies. The coatings used on automobile steel is thermoset as well. The clear coatings will eventually crack and peel on the edges of exterior applications. On interior applications the lifespan and performance of the coatings is significantly better. The environment is less severe, and the ultraviolet radiation is negligible (Figure 8.21).

The clear protective coatings fall into several categories. Each category has shown various levels of success in protecting the surfaces of steel. In each instance it is critically important to ensure that the surface is clean and free of oils and, in the case of the weathering steels and blackened steels, all ionic reactants that may be present on the surface. Rinse the surface with deionized water to pull the free ions from the surface. Thoroughly dry the surface, possibly by warming the surface in a dry, nonhumid environment and follow with wiping down with isopropyl alcohol. Once dry, apply the coating.

The clear lacquers are essentially paints. Incralac™, a clear acrylic coating used frequently to protect copper alloys from corrosion and oxidation, has shown success on steel surfaces. Other acrylics and polyurethanes that can be used on the steels as well.

The hydrocarbon-type drying oils are used in industry to protect steel surfaces exposed to rigorous environments. They dry to a hard, clear coating but lack some of the even layering you get with additives in lacquers. There are several on the market sold as hydrophobic and oleophobic coatings. On weathering steels, these coatings will darken the surface.

The UV-stable varnishes, (UVS varnishes) are used by museums to provide a reversable, long-lasting, ultraviolet stable varnish. They have good leveling characteristics and can be layered.

Waxes are common and they can be used in tandem with the clear lacquers and the varnishes. They will age and yellow over time, usually three to five years. On exterior applications, the steel will get quite warm and these can decay more rapidly. They need to have a scheduled replacement and refurbishment for these to succeed.

The nano-coatings are newer to the marketplace and are showing promise. These coatings are often silicon based and have special hydrophobic characteristics to assist in shedding moisture from the surface. Sol-gels and other siloxane-based coatings are in development across industries to provide oxidation protection to various metals.

For steels, lanolin-based coatings are good, temporary hydrophobic coatings. They have shown wide use in industries as a wipe-on coating to protect steels. They do exude an odor. These coatings were discovered anecdotally, it is said, by sheep herders. They noticed old steel vehicles that had been left on the farm rusted extensively except where the sheep would rub against the surface. These areas remained rust free. There are several lanolin-based coatings that will protect the steel from corroding. They darken the surfaces of weathering steels.

Choji oil is a Japanese oil used to protect high-quality steel sword blades. This oil is extracted from cloves and combined with a light mineral oil. The steel absorbs the oil and dries. There is an initial odor with this treatment.

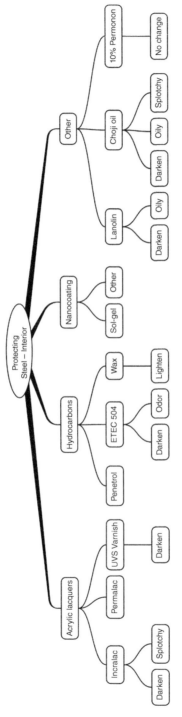

FIGURE 8.21 Examples of various surface coatings for steel.

FIGURE 8.22 "Book of Knowledge" sculptures made of weathering steel and stainless steel.
Source: Artist Steven Woodward.

Steels for all their toughness can be an elegant design choice (Figure 8.22). Knowledge and good practice are the keys to success. Steels have been used for centuries and are seeing a resurgence in art and architecture in the twenty-first century.
‘

Terms

Allotropic	Different crystal structures within the same steel.
Anisotropic	Having different mechanical behaviors when acted on in one direction as opposed to a perpendicular direction.
Annealing	Steel heated to the austenite range, then slowly cooled to soften and remove hardness.
Austenite	One of the three allotropes of iron. Known as gamma-phase iron or Ý-Fe. Form of face centered cubic structure of iron crystal. Forms at $910\,°C$.
Bessemer converter	Open hearth, electric arc furnace converts pig iron to steel.
Blast furnace	Towering furnace made of brick. Air is blasted through the lower section to reduce ore to pig iron.
Capped	Similar to rimmed steel. A cap is placed on the ingot mold to solidify the top metal and stopping the rim development. Surface quality is similar to rimmed steel.
Carburized	The absorption of carbon into a steel surface by heating the steel to an elevated temperature in the presence of a carbon source. This produces a carbon-enriched layer into the surface of the steel in variable concentrations as the carbon is absorbed. This greatly hardens the surface. It is a form of case hardening.
Case hardening	A process or series of similar processes where elements are absorbed and diffused into the surface of steel. Carbon and nitrogen are the normal elements that get absorbed into the surface layer. The mechanical properties of the surface are altered by this absorption.
Cementite	Iron carbide. Hard and brittle; Fe_3C.

Cold-rolled steel	A mill product whereby the coil of steel is passed through a set of cold rolls that reduce a hot-rolled steel coil and produce a temper into the metal. The surface is usually smoother and flatter than hot-rolled steel.
Commercial quality, CQ	Older standard designation for mild carbon steels that sets a minimum-quality level. The steel is ductile and formable. This is now referred to as CS, or commercial steel.
Commercial steel, CS	Standard designation for mild carbon steels that sets a minimum quality level. The steel is ductile and formable.
Continuous casting	A technique whereby the steel is cast into slabs or billets in a continuous fashion. The cast slab or billet travels to a hot-rolling station where it is reduced. All of this is performed on a continuous line.
Corten	A trademark of US Steel. This is the high-strength low alloy steel today known as weathering steel. Also spelled COR-TEN®. 'Cor' for corrosion resistance. 'Ten' for high tensile strength.
Damascene steel	A type of forged steel from ancient times that used wootz steel imported from ancient India and Tamraparni. The art of making this type of steel is lost with pattern-welded steel being the closest replication of the end effect.
Darkened steel	Carbon steel that has been treated with bluing or darkening treatments. These can be cold or hot treatments that develop a black surface layer of magnetite, Fe_3O_4, or a layer of copper selenide. Another form of bluing or gun bluing.
Deep drawing steel, DDS	Standard designation for carbon steel with refined grain characteristics, capable of forming and drawing operations.
Ductile iron	A cast iron that has been treated to generate nodules or sheroids of graphite. The treatment is usually the addition of another element, usually magnesium or cerium, to induce the formation of free graphite. Also known as nodular cast iron.
Extra deep drawing steel, EDDS	Standard designation for carbon steel with special characteristics designed by the end user to enable severe forming operations. The steel is very ductile.
Ferrite	One of the three allotropic forms of iron. Stable at room temperature. Known also as alpha iron or $\alpha - Fe$.
Forge welding	Heating two pieces of metal together and joining them under high pressure by means of a forge.
Grain	A crystal within a polycrystalline surface. The crystal is made up of an arrangement of atoms.

Grain boundary	The narrow line where one crystal ends and another begins.
Gray cast iron	When fractured, the gray flakes of graphite are visible. Good toughness.
Gun bluing	A process whereby carbon steel is treated to develop a partial corrosive resistance barrier on the surface. Used extensively on steel firearms, gun bluing forms a layer of magnetite, Fe_3O_4 over the surface.
Hardening	Raising the temperature to form austenite then rapidly cooling.
Hot-rolled steel	A slab or coil of thick steel is rolled while still hot and reduced in thickness but elongated and widened.
HSLA	High-strength, low-alloy steels. See also weathering steels.
Isotropic	Having the same mechanical behavior in all directions.
Killed	Steel that has been deoxidized by treating with a strong deoxidizing element, usually aluminum or silicon to remove the dissolved oxygen. This is done in the molten state and the subsequent steel has no carbon and oxygen reaction. It does not 'boil' as the carbon monoxide is released, instead it lays still – thus, the term *killed*.
Maraging	Refers to a special group of nickel-bearing steels that have been precipitation hardened to improve strength and hardness.
Martensite	Hard, needle-like phase of iron. Forms when steel is rapidly cooled from the austenite state and carbon gets trapped and forms cementite.
Mild steel	Low carbon steel. 0.05–0.06% carbon.
Nodular cast iron	Standard designation for mild carbon steels that sets a minimum quality level. The steel is ductile and formable.
Normalization	Heating to the austenitic range followed by cooling to produce small grain structure.
Pattern-welded steel	The welding together of several different compositions of steel by forge welding them together.
Pearlite	Mixture of ferrite and cementite. Lower melting point than the constituents.
Preweathered steel	Term given to describe the artificial weathering of the group of high-strength, low-alloy steels (HSLA) commonly referred to as weathering steels. The artificial weathering is performed in a controlled environment in order to achieve the protective oxide and pleasing color before in advance.
Rimmed	Carbon steel that releases carbon monoxide while solidifying that leaves a case or rim of steel free of voids as the carbon monoxide bubbles out. This produces good surface quality on the sheet or plate of steel produced from the rim portion of the steel.
Scale	Adherent glassy substance that forms on the surface of cooling steel.

Semi-killed	Similar to killed steel but with some carbon monoxide development. Reduces shrinkage during solidification.
Slag	Waste product remnant from blast furnace operation. Rock-like consistency.
Solanum	A trademark of Zahner. Used to describe the preweathering process of high-strength, low-alloy steels.
Structural steel, SS	Standard designation for carbon steels with specific mechanical properties used in areas subject to structural requirements established by engineering design criteria.
Sulfurized	Carbon steel where sulfur is added back to improve machining.
Tempering	Raising the temperature to just below austenite formation to remove hardness.
Tool steel	High-carbon steel. 0.06–1.7% carbon.
Watered steel	The term given to damascene steel due to the patterns visible on the surface. Watered steel was the term used more often in Europe. While Damascus or damascene was used in the east.
Weathering steel	Term given to a group of high-strength, low-alloy steels (HSLA) that develop a protective oxide when exposed to the atmosphere. The oxide is corrosion resistant and provides a pleasing color once fully developed. Also known as Corten or COR-TEN®, the trademark of US Steel, the original inventor.
White cast iron	Cast iron that, when fractured, has a white appearance due to carbon integrated into the iron.
Wootz steel	High-carbon steel made in crucibles containing porous iron ore along with wood or charcoal. The technique has been lost over the years. This was the steel used to make the famous damascene steel.
Wrought iron	High-purity iron with slag inclusions.
Wrought steel	Pure ferritic iron with slag inclusions. Less than 0.04% carbon.

Comparative Attributes of Metals

Attribute	Steel	Stainless steel	Aluminum	Copper alloys	Zinc
Cost – relative	Low	High	Medium	High	High
Recyclability	Excellent	Excellent	Excellent	Excellent	Excellent
Strength	High	Medium to high	Low to medium	Low to medium	Low
Hardness	High	High	Low	Low	Low
Ease of casting	Good	Fair	Good	Excellent	Excellent
Formability	Good	Good	Excellent	Excellent	Good
Tensile strength	High	Good	Fair	Fair	Poor
Ductility	Good	Good	Excellent	Excellent	Good
Vibration dampening	Good	Good	Excellent	Excellent	Good
Weldability	Excellent	Good	Good	Excellent	Good
Machability	Good	Fair	Excellent	Good	Excellent
Stamp	Excellent	Good	Excellent	Excellent	Good
Extrudability	Fair	Fair	Excellent	Good	Good
Thermal expansion 10^{-6} m/m°C	6.8	9.6	13.1	9.8	19

Attribute	Steel	Stainless steel	Aluminum	Copper alloys	Zinc
Electrical conductivity S/m at 20°C	1×10^7	1.45×10^6	3.77×10^7	5.96×10^7	1.69×10^7
Corrosion resistance	Poor to excellent (weathering steel)	Excellent	Excellent	Excellent	Good
Color (natural and oxide development)	Gray-blue, black, reddish brown	Silver, blue, gold, maroon, green, black, bronze	Silver, black, dye colors	Salmon red, yellows, golds, pink, green, black, red, gray, silver	Gray blue, black, yellow brown, white

APPENDIX C

Alloy Cross References

The following table is an attempt to arrive at some understanding of the cross-referencing of alloys from one specification to another. This represents only an approximation of the UNS alloy number to the AISI alloy number to the European designation.

UNS Alloy	AISI / ASTM	EUROPEAN	
G10060	1006	EN 10016	C7D
G10080	1008	EN 10016	C9D
G10100	1010	EN 10016	C10D
K02600	A36	EN 10025	
G10180	1018	EN 10016	C18D
G10200	1020	EN 10016	C20D
K11430	A588	EN10113	S355JOW
K12043	A588	EN10113	S355J2P
G41300	4130	EN 10083	25CrMo 4

Mechanical properties and alloying constituents are similar but not exact.

These are the alloys discussed in this book. The standards listed in this book are an attempt to designate what may be important to a designer using the steels as an aesthetic material rather than what a structural engineer might need. In the event that mechanical properties are important criteria, then the mechanical properties of the particular steel must be examined more precisely.

APPENDIX D

Abrasive Preparation of Steels

(SSPC: Society for Protective Coatings, www.sspc.org)
(NACE: NACE International. National Association of Corrosion Engineers, www.nace.org)

Designation	Basic Procedure
SSPC-SP1 Solvent cleaning	Remove oil, grease, soils, with solvents.
SSPC-SP2 Hand tool cleaning	Remove loose rust and paint particles by wire brush, hand sanding, or other manual device.
SSPC-SP3 Power tool cleaning	Use power sanders, grinders, or descalers to remove all loose scale and foreign matter.
SSPC-SP5/NACE 1 White metal blast cleaning	Surface is free of all oils, oxides, scale, and old coatings.
SSPC-SP6/NACE 3 Commercial blast cleaning	No more than 33% of a 3 in. square area of the surface can show any rust, oxides, or old coatings.
SSPC-SP7/NACE4 Brush-off blast cleaning	Some mill scale can be on the surface if it is very tightly adhered and cannot be scraped off.
SSPC-SP10/NAC2 Near white blast cleaning	No more than 5% of a 3 in. square area of the surface can show any rust, oxides, or old coatings.
SSPC-SP11 Power tool cleaning to bare metal	No visible dirt, mill scale, paint, or corrosion products. If pitted, surface the base of the small pits can still have particles.
SSPC-SP14/NACE8 Industrial blast cleaning	Clean by blasting to remove all loose particles over 90% of the surface as long as the 10% remaining is evenly distributed.

Designation

SSPC-SP WJ-1/NACE WJ-1 Surface preparation and cleaning by high-pressure waterjet

SSPC-SP WJ-2/NACE WJ-2 Surface preparation and cleaning by high-pressure waterjet

SSPC-SP WJ-3/NACE WJ-3 Surface preparation and cleaning by high-pressure waterjet

SSPC-SP WJ-4/NACE WJ-4 Surface preparation and cleaning by high-pressure waterjet

SSPC-SP 5/NACE WAB-1 Surface preparation and cleaning by wet abrasive blast cleaning

SSPC-SP 6/NACE WAB-3 Surface preparation and cleaning by wet abrasive blast cleaning

SSPC-SP 7/NACE WAB-4 Surface preparation and cleaning by wet abrasive blast cleaning

SSPC-SP 10/NACE WAB-2 Surface preparation and cleaning by wet abrasive blast cleaning

Basic Procedure

Preparation of a steel surface using high-pressure waterjet. No loose rust, scale, or previous coatings. Removal is down to the base metal.

Preparation of a steel surface using high-pressure waterjet. Very thorough removal of all loose rust, scale, or previous coatings. Small amounts of the oxides can be present.

Removal of all loose particles of mill scale, rust, coatings, oil, and grease. Some degree of tightly adhering films can be present.

Blasting with waterjet to remove most of the loose particles as practical. The surface will be uneven and mottled.

Removal of all surface oils, paint, oxides, and scale by wet abrasive blast. No signs of stains from these coatings will remain.

No more than 33% of a 3 in. square area of the surface can show any rust, oxides, or old coatings.

Brush off surface with wet abrasive blast to remove loose paint and oxides. Some will remain. Leave a roughened surface.

Near white metal wet abrasive blast leaving the surface clean and free of all scale, rust, oxide, coatings, and oils, but the stains from these coatings may remain.

There are three approaches used in the preparation of carbon steels:

1. Dry blast
2. Waterjet blast
3. Water blast with abrasive

For steels, it is important when using the wet systems to arrive at an understanding of how much flash rust or surface uniform oxidation that appears quickly after cleaning will be allowed. Because these systems introduce water, the surfaces can develop surface rust quickly.

APPENDIX E

Specification Information

This appendix lists the alloy type and basic quality steps to include in the specification. When working with steels, it is important to realize that art and architecture are not the market many of the supply houses and mill producers consider. Therefore, it is necessary to go more in depth in what you expect out of the metal surface and work with a fabricator to achieve the end results desired.

Appendix E covers six steel wrought forms commonly used in art and architecture.

E.1.   Natural steel surface produced at the mill for interior applications

E.2.   Weathering steel expected to weather naturally for exterior applications

E.3.   Preweathered steel for exterior applications

E.4.   Preweathered steel for interior applications

E.5.   Darkened steel for interior applications

E.6.   Darkened and mottled steel for interior applications

E.1. NATURAL STEEL SURFACE

USE: Interior

SUBJECT OF SPECIFICATION: NATURAL STEEL

	ASTM STANDARDS
METAL: Low-Carbon Steel	**ASTM A29/A29M**
UNS ALLOY #: G10080	**ASTM A505-16**
G10010	**ASTM A659/A659M**
	ASTM A1008/A1008M

Thickness: *It is recommended to use 2 mm (0.079 inch)*

SURFACE FINISH DESCRIPTION

The surface of the steel will be natural, mill finish steel. The surface will receive a clear protective coating. The clear coating will be matte and free of 'orange peel,' runs, or other coating issues. The mill surface will have variations of gray to blue-gray tones and possess a linear character.

QUALITY ASSURANCE

Prototypes and Samples:

Produce representative samples, minimum of three 1 M square to act as target color and gloss tones. The clear coating to be used on the project will be applied to these representations. These are target finishes representing the desired final finish. As the actual metal for the project is being processed, new samples will be produced and compared to these representations.

> *Note: Hot-rolled plate and sheet will have a slightly rougher surface than cold-rolled sheet. Be certain you consider both as the appearance will vary.*

Metal:

The metal will be received at the factory for processing free of major scratches or gouges. There will be no dents, oxides, scale, or scars from scale removal. The metal should be oiled and pickled from the supplier house or mill.

Factory:

After inspecting the metal for scratches and mars, the surface will be degreased and prepared for clear coating. The surface of the steel will maintain the quality provided from the mill and approved by the designer. No processing of the surface such as polishing or grinding should occur.

Remove all fingerprints and handprints before coating.

Once clean, apply the clear coating to protect the surface from oxidizing.

The clear coat shall be compatible with the steel and applied soon after all fabrication has occurred.

The clear coat shall be capable of providing a minimum of 7 years before recoating is required unless previously approved by the designer.

Handling:

Do not allow moist wood or packaging material to contact the surface for any length of time. Use caution in handling the surface to avoid marring the surface or damaging the clear coating. Store the steel in a clean, dry location. Do not stack or place anything on the face of the metal surface.

E.2. WEATHERING STEEL SURFACE – NATURAL WEATHERING

USE: Exterior

SUBJECT OF SPECIFICATION: WEATHERED STEEL–NATURAL WEATHERING

METAL:	ASTM STANDARDS
High-Strength, Low-Alloy Carbon Steel	**ASTM A588/A588M**
UNS ALLOY #:K11430	**ASTM A606/606M**
K12043	**ASTM A242/242M**

Thickness: *Depends on use. No less than 2 mm for sheet.*

SURFACE FINISH DESCRIPTION

The surface of the steel will be clean and free of all oils and scale. The surface will be even with no mechanical processes performed after cold or hot rolling. Some initial surface oxide is expected.

QUALITY ASSURANCE

Prototypes and Samples:

Produce representative samples, from one to three for verification of thickness and to represent the surface as it will be installed.

> *Note: Hot-rolled plate and sheet will have a slightly rougher surface than cold-rolled sheet. Be certain you see all thicknesses. Hot-rolled and cold-rolled steel will develop different oxide levels.*

Metal:

The metal will be received at the factory for processing free of major scratches or gouges. There will be no dents, heavy scale or scars from scale removal. Banding should be avoided.

Factory:

The metal will be degreased if needed.

Handling:

Do not allow moist wood or packaging material to contact the surface for any length of time.

Use caution in handling the surface to avoid marring the surface.

Store the preweathered steel plates in a clean, dry location. Do not stack or place anything on the face of the metal surface. Allow the surface to get air.

If allowed to get wet, avoid runs by storing the material vertically. Do not allow water to pond on the surface.

Note: You should expect the metal to undergo changes depending on exposure. Moisture is the catalyst to the development of the oxide. It is critical to get the surface installed without large stains from oils or moisture that has pooled on the metal.

E.3. PREWEATHERED STEEL

SUBJECT OF SPECIFICATION: PREWEATHERED STEEL

USE: Exterior

METAL:	ASTM STANDARDS
High Strength Low Alloy Carbon Steel	**ASTM A588/A588M**
UNS ALLOY #: K11430	**ASTM A606/606M**
K12043	**ASTM A242/242M**

Thickness: *Depends on use. No less than 2 mm for sheet.*

SURFACE FINISH DESCRIPTION

The surface of the steel will have a stable developed oxide. The color induced by the oxide will be consistent across the exposed surface. The color shall be a rich reddish brown covering 95% of the area with some blackened regions and some lighter orange regions.

No visible streaks from the wetting process.

QUALITY ASSURANCE

Prototypes and Samples:

Produce representative samples, minimum of three, 1 M square, in the thickness to be used and with the finish fully produced. Once approved, these will become the target surfaces to be used as controls for the project.

Note: Hot-rolled plate and sheet will have a slightly rougher surface than cold-rolled sheet. Be certain you see all thicknesses. Hot-rolled and cold-rolled steel will develop different oxide levels.

Metal:

The metal will be received at the factory for processing free of major scratches or gouges.

There will be no dents, heavy scale, or scars from scale removal. Banding should be avoided.

Factory:

The metal will be degreased and all mill oil will be removed before processing.

The metal will be blasted to SSPC - SP5/NACE 1, white metal blast, to remove the surface oxides. *It is preferred to use an automated blasting system to eliminate mottling.*

The metal surface will be activated with an appropriate oxidizing solution with all dangerous effluent capture and properly disposed of so no harm will come to the environment or personnel. Process the metal in a controlled environment where humidity and temperature are maintained throughout the process as the metal surface develops the oxide. Monitor the development of the oxide and the stabillity of the oxide. Final stages of the oxide development should have little to no residue come off on hands or when brushed with a cotton rag.

Handling:

Once the oxide is generated to sufficient levels, exercise care in the handling and the packaging of the metal surface. If the surface is allowed to get wet and a stain develops, this surface will need to be reprocessed to remove the stain. Do not allow moist wood or packaging material to contact the surface for any length of time.

Caution in handling the surface to avoid marring the initial fragile oxide. If this occurs, re-wet the surface and bring the oxide back by renewing the wet dry cycles.

Surfaces with visible scratches, gouges, or mars, when viewed from a distance of 4 meters should be reworked or replaced until they are acceptable.

Store the preweathered steel plates in a clean dry location. Do not stack or place anything on the face of the metal surface. Allow the surface to get air.

Dust development during storage is common. Rinse the surface with clean water and allow to dry. Do not allow water to pool on the surface and avoid runs. A light misting is preferred.

E.4. PREWEATHERED STEEL FOR INTERIOR APPLICATIONS

USE: Interior

SUBJECT OF SPECIFICATION: PREWEATHERED STEEL

METAL:	ASTM STANDARDS
High-Strength Low-Alloy Carbon Steel	**ASTM A588/A588M**
UNS ALLOY #: K11430	**ASTM A606/606M**
K12043	**ASTM A242/242M**

Thickness: *Depends on use. No less than 0.8 mm for sheet.*

SURFACE FINISH DESCRIPTION

The surface of the steel will have a stable developed oxide. The color induced by the oxide will be consistent across the exposed surface. The color shall be a rich reddish brown covering 95% of the area with some blackened regions and some lighter orange regions. The surface will be slighty burnished. The surface will be smooth and no apparent dust or loose oxide.

QUALITY ASSURANCE

Prototypes and Samples:

Produce representative samples, minimum of three, 1 M square, in the thickness to be used and with the finish fully produced. Once approved, these will become the target surfaces to be used as controls for the project.

Note: Hot-rolled plate and sheet will have a slightly rougher surface than cold-rolled sheet. Be certain you see all thicknesses. Hot-rolled and cold-rolled steel will develop different oxide levels.

Metal:

The metal will be received at the factory for processing free of major scratches or gouges.
There will be no dents, heavy scale, or scars from scale removal. Banding should be avoided.

Factory:

The metal will be degreased and all mill oil will be removed before processing.
The metal will be blasted to SSPC - SP5/NACE 1, white metal blast, to remove the surface oxides.
It is preferred to use an automated blasting system to eliminate mottling.

The metal surface will be activated with an appropriate oxidizing solution with all dangerous effluent capture and properly disposed of so no harm will come to the environment or personnel.

Process the metal in a controlled environment where humidity and temperature are maintained throughout the process as the metal surface develops the oxide. Monitor the development of the oxide and the stabilility of the oxide. Work the surface to remove all excess oxide that has not bonded with the surface. This will slightly burnish the surface. All excess oxide will be removed. Final stages of the oxide development should have no residue when brushed with a cotton rag.

Handling:

Once the oxide is generated to sufficient levels, exercise care in the handling and the packaging of the metal surface. If the surface is allowed to get wet and a stain develops, this surface will need to be reprocessed to remove the stain. Do not allow moist wood or packaging material to contact the surface for any length of time.

Use caution in handling the surface to avoid marring the surface. For interior usage, the surface must be worked in the factory to eliminate all remnants of the fragile oxide.

Surfaces with visible streaks, scratches, gouges, or mars, will not be acceptable.

Store the preweathered steel plates in a clean dry location. Do not stack or place anything on the face of the metal surface. Allow the surface to get air.

E.5. DARKENED STEEL SURFACE – CONSISTENT/NONMARBLED

USE: Interior

SUBJECT OF SPECIFICATION: DARKENED STEEL – BLACK

METAL:	ASTM STANDARDS
Low-Carbon Steel	**ASTM A29/A29M**
UNS ALLOY #: G10080	**ASTMA505-16**
G10010	**ASTM A659/A659M**
	ASTM A1008/A1008M

Thickness: *It is recommended to use 2 mm (0.079 Inch)*

SURFACE FINISH DESCRIPTION

The surface of the steel will be matte black in appearance with a clear protective coating applied. The clear coating will be matte and free of 'orange peel,' runs, or other coating issues.

Variations from panel to panel will be visible due to the fact this is a chemically induced surface and as such has numerous variables in the process.

QUALITY ASSURANCE

Prototypes and Samples:

Produce representative samples, minimum of three 1 M square to act as target color and gloss tones. The clear coating to be used on the project will be applied to these representations. These are target finishes representing the desired final finish. As the actual metal for the project is being processed, new samples will be produced and compared to these representations.

> *Note: Hot-rolled plate and sheet will have a slightly rougher surface than cold-rolled sheet. Be certain you consider both as the appearance will vary.*

Metal:

The metal will be received at the factory for processing free of major scratches or gouges. There will be no dents, oxides, scale, or scars from scale removal. The metal should be oiled and pickled from the supplier house or mill.

Factory:

After inspecting the metal for scratches and mars, the surface will be degreased and prepared for darkening. The surface of the steel will maintain the quality provided from the mill and approved by the designer. No processing of the surface such as polishing or grinding should occur. Apply either a hot blackening process or a cold darkening process identical to the approved proto-type surfaces. If the hot blackening process is used, protect all personnel in accordance with OSHA and ensure the facility meets all regulatory criteria. Rinse and dry the surface before final clear coat. For both processes, use a facility versed in the methods of proper disposal of effluent. All effluent will be captured and recycled or properly disposed of in accordance with state and local laws. Proof of this should be submitted with the prototypes and approved before beginning. Immediately after darkening, apply the clear coating to protect the surface from oxidizing. The clear coat shall be compatible with the blackening treatment and applied soon after darkening. The clear coat shall be capable of providing a minimum of 7 years before recoating is required unless previously approved by the designer.

Handling:

Do not allow moist wood or packaging material to contact the surface for any length of time. Caution in handling the surface to avoid marring the surface or damaging the clear coating. Store the darkened steel in a clean, dry location. Do not stack or place anything on the face of the metal surface.

E.6. DARKENED STEEL SURFACE – MARBLED

USE: Interior

SUBJECT OF SPECIFICATION: DARKENED STEEL – VARIGATED/MARBLED

METAL: Low-Carbon Steel	**ASTM STANDARDS**
UNS ALLOY #: G10080	**ASTM A29/A29M**
G10010	**ASTM A505-16**
	ASTM A659/A659M
	ASTM A1008/A1008M

Thickness: *It is recommended to use 2 mm (0.079 inch)*

SURFACE FINISH DESCRIPTION

The surface of the steel will be custom marbled black in appearance. The surface will receive a clear protective coating. The clear coating will be matte and free of 'orange peel,' runs, or other coating issues. The marbling tones are highly variable and are not an exacting process. It is a natural chemical reaction that develops dark and light tones across the surface of steel.

QUALITY ASSURANCE

Prototypes and Samples:

Produce representative samples, minimum of three 1 M square to act as target color and gloss tones. The clear coating to be used on the project will be applied to these representations. These are target finishes representing the desired final finish. As the actual metal for the project is being processed, new samples will be produced and compared to these representations.

Note: Hot-rolled plate and sheet will have a slightly rougher surface than cold-rolled sheet. Be certain you consider both as the appearance will vary.

METAL:

The metal will be received at the factory for processing free of major scratches or gouges.
　　There will be no dents, oxides, scale, or scars from scale removal. The metal should be oiled and pickled from the supplier house or mil.

Factory:

After inspecting the metal for scratches and mars, the surface will be degreased and prepared for darkening. The surface of the steel will maintain the quality provided from the mill and approved by the designer. No processing of the surface such as polishing or grinding should occur. The marbling is applied using a cold darkening process identical to the approved prototype surfaces. Rinse and dry the surface before final clear coat.

Use a facility versed in the methods of proper disposal of effluent generated from the process. All effluent will be captured and recycled or properly disposed of in accordance with state and local laws.

Proof of this should be submitted with the prototypes and approved before beginning.

Immediately after darkening, apply the clear coating to protect the surface from oxidizing.

The clear coat shall be compatible with the blackening treatment and applied soon after darkening.

The clear coat shall be capable of providing a minimum of 7 years before recoating is required unless previously approved by the designer.

Handling:

Do not allow moist wood or packaging material to contact the surface for any length of time.

Use caution in handling the surface to avoid marring the surface or damaging the clear coating.

Store the darkened steel in a clean, dry location. Do not stack or place anything on the face of the metal surface.

A Few Relevant specifications from the ASTM International – (Not a Complete List)

A6/A6M	General requirements	Rolled structural steel bars, plates, shapes, and sheet piling
A27/A27M	Standard specification	Steel castings, carbon, for general application
A29/A29M	General requirements	Steel bars, carbon and alloy, hot-wrought
A36/A36M	Standard specification	Carbon structural steel
A48/A48M	Standard specification	Gray iron castings
A53/A53M	Standard specification	Pipe, steel, black and hot-dipped, zinc-coated, welded and seamless
A108	Standard specification	Steel bar, carbon and alloy, cold-finished
A90/A90M	Standard test method	For weight of coating on iron and steel articles with zinc or zinc-alloy coatings
A109/A109M	Standard specification	Steel, strip, carbon (0.25 maximum percent) cold-rolled
A123/A123M	Standard specification	Zinc (hot-dip galvanized) coatings on iron and steel products
A148/A148M	Standard specification	Steel castings, high strength, for structural purposes
A227/A227M	Standard specification	Steel wire, cold-drawn for mechanical springs
AA228/A228M	Standard specification	Steel wire, for music string quality
A242/A242M	Standard specification	For high-strength, low-alloy structural steel

ASTM F3125	Standard specification	For high-strength structural bolts and assemblies, steel and alloy steel
A283/A283M	Standard specification	For low and intermediate tensile strength carbon steel plates
A322	Standard specification	For steel bars, alloy standard grades
A370	Standard test method	Definitions for mechanical testing of steel products
A385/A385M	Standard practice	For providing high-quality zinc coatings (hot-dip)
A424/A424M	Standard specification	For steel sheet for porcelain enameling
A434/A434M	Standard specification	For steel bars, alloy, hot-wrought or cold-finished, quenched and tempered
A463/A463M	Standard specification	For steel sheet, aluminum-coated, by the hot-dip method
A505	Standard specification	For steel sheet and strip, alloy, hot-rolled and cold-rolled general requirements
A506	Standard specification	For alloy and structural alloy steel, sheet and strip, hot-rolled and cold-rolled
A568/A568M	Standard specification	For steel, sheet, carbon, structural, and high-strength, low-alloy, hot-rolled and cold-rolled general requirements
A588/A588M	Standard specification	For high-strength, low-alloy structural steel, up to 50 ksi (345 Mpa) minimum yield point, with atmospheric corrosion resistance
A600	Standard specification	For tool steel high speed
A606 /A606M	Standard specification	For steel sheet and strip, high-strength, low-alloy, hot-rolled and cold-rolled, with improved atmospheric corrosion resistance
A653/A653M	Standard specification	For steel sheet, zinc-coated (galvanized) or zinc iron alloy-coated (galvannealed) by the hot-dip method
A656/A656M	Standard specification	For hot-rolled structural steel, high-strength, low-alloy plate with improved formability
A659/A659M	Standard specification	For commercial steel (CS), sheet and strip, carbon (0.25 max) hot rolled
679/679M	Standard specification	For steel wire, high tensile strength, cold drawn
A1008/A1008M	Standard specification	For commercial steel, sheet, cold-rolled, carbon, structural, high-strength, low-alloy and high-strength, low-alloy with improved formability
A1018/A1018M	Standard specification	For steel, sheet and strip, heavy thickness coils, hot-rolled, carbon, structural, high-strength, low-alloy, Columbium or Vanadium, and high-strength, low-alloy with improved formability

Hot-Dipped Galvanizing Specification

Hot-Dip Galvanized Grade	Minimum μm/side	Minimum mils/side
35	35	1.38
45	45	1.77
50	50	1.97
55	55	2.17
60	60	2.36
65	65	2.56
75	75	2.95
80	80	3.15
85	85	3.35
100	100	3.94

For the hot-dip, galvanized coating applied by immersing individual parts in the molten zinc bath. The thickness is impacted by the steel thickness.

Continuous Sheet Galvanizing	Approx. µm/side	Approx. mil/side
G60	13.7	0.54
G90	20.6	0.81
G115	26.3	1.04
G140	32.0	1.26
G165	37.7	1.49
G185	42.3	1.67
G210	48.0	1.89
G235	53.7	2.12
G300	68.6	2.70
G360	82.3	3.24

For the continuous sheet, hot dipped galvanizing, the value following the letter G corresponds to the total weight of zinc applied to both sides of the steel sheet.

Index

Note: Page references in *italics* refer to figures and tables.